Graphics Concepts
with SolidWorks 2000®

Richard M. Lueptow
Northwestern University

Michael Minbiole

Prentice Hall
Upper Saddle River, NJ 07458

Library of Congress Cataloging-in-Publication Data

Lueptow, Richard M.
    Graphics concepts with SolidWorks / Richard M. Lueptow, Michael Minbiole.
      p.   cm. — (ESource—the Prentice Hall engineering source)
    Includes bibliographical reference and index.
    ISBN 0–13–014155–0
    1. Computer graphics. 2. SolidWorks. 3.Engineering models. 4. Computer-aided design.
       I. Minbiole, Michael. II. Title. III. Series.

T385.L84 2000
620'.0042'02855369—dc21
                                                          00.060684

Vice-president of editorial development, ECS: **MARCIA HORTON**
Acquisitions editor: **ERIC SVENDSEN**
Associate editor: **JOE RUSSO**
Vice-president of production and manufacturing: **DAVID W. RICCARDI**
Executive managing editor: **VINCE O' BRIEN**
Managing editor: **DAVID A. GEORGE**
Editorial/production supervisor: **LAKSHMI BALASUBRAMANIAN**
Cover director: **JAYNE CONTE**
Manufacturing buyer: **PAT BROWN**
Editorial assistant: **KRISTEN BLANCO**
Market manager: **DANNY HOYT**

 © 2000 by Prentice-Hall, Inc.
Upper Saddle River, New Jersey 07458

All rights reserved. No part of this book may be
reproduced, in any form or by any means,
without permission in writing from the publisher.

The author and publisher of this book have used their best efforts in
preparing this book. These efforts include the development, research,
and testing of the theories to determine their effectiveness.
Printed in the United States of America.

10 9 8 7 6 5 4 3 2 1

ISBN 0-13-014155-0

Prentice-Hall International (UK) Limited, *London*
Prentice-Hall of Australia Pty. Limited, *Sydney*
Prentice-Hall Canada, Inc., Toronto
Prentice-Hall Hispanoamericana, S.A., *Mexico*
Prentice-Hall of India Private Limited, *New Delhi*
Prentice-Hall of Japan, Inc., *Tokyo*
Pearson Education (Singapore) Pte. Ltd., *Singapore*
Editoria Prentice-Hall do Brasil, Ltda., *Rio de Janeiro*

About ESource

ESource—The Prentice Hall Engineering Source
—www.prenhall.com/esource

ESource—The Prentice Hall Engineering Source gives professors the power to harness the full potential of their text and their first-year engineering course. More than just a collection of books, ESource is a unique publishing system revolving around the ESource website—www.prenhall.com/esource. ESource enables you to put your stamp on your book just as you do your course. It lets you:

Control You choose exactly what chapter or sections are in your book and in what order they appear. Of course, you can choose the entire book if you'd like and stay with the authors' original order.

Optimize Get the most from your book and your course. ESource lets you produce the optimal text for your students needs.

Customize You can add your own material anywhere in your text's presentation, and your final product will arrive at your bookstore as a professionally formatted text.

ESource ACCESS

Starting in the fall of 2000, professors who choose to bundle two or more texts from the ESource series for their class, or use an ESource custom book will be providing their students with complete access to the library of ESource content. All bundles and custom books will come with a student password that gives web ESource ACCESS to all information on the site. This passcode is free and is valid for one year after initial log-on. We've designed ESource ACCESS to provides students a flexible, searchable, on-line resource.

ESource Content

All the content in ESource was written by educators specifically for freshman/first-year students. Authors tried to strike a balanced level of presentation, an approach that was neither formulaic nor trivial, and one that did not focus too heavily on advanced topics that most introductory students do not encounter until later classes. Because many professors do not have extensive time to cover these topics in the classroom, authors prepared each text with the idea that many students would use it for self-instruction and independent study. Students should be able to use this content to learn the software tool or subject on their own.

While authors had the freedom to write texts in a style appropriate to their particular subject, all followed certain guidelines created to promote a consistency that makes students comfortable. Namely, every chapter opens with a clear set of **Objectives**, includes **Practice Boxes** throughout the chapter, and ends with a number of **Problems**, and a list of **Key Terms**. **Applications Boxes** are spread throughout the book with the intent of giving students a real-world perspective of engineering. **Success Boxes** provide the student with advice about college study skills, and help students avoid the common pitfalls of first-year students. In addition, this series contains an entire book titled ***Engineering Success*** by Peter Schiavone of the University of Alberta intended to expose students quickly to what it takes to be an engineering student.

Creating Your Book

Using ESource is simple. You preview the content either on-line or through examination copies of the books you can request on-line, from your PH sales rep, or by calling 1-800-526-0485. Create an on-line outline of the content you want, in the order you want, using ESource's simple interface. Either type or cut and paste your own material and insert it into the text flow. You can preview the overall organization of the text you've created at anytime (please note, since this preview is immediate, it comes unformatted.), then press another button and receive an order number for your own custom book. If you are not ready to order, do nothing—ESource will save your work. You can come back at any time and change, re-arrange, or add more material to your creation. You are in control. Once you're finished and you have an ISBN, give it to your bookstore and your book will arrive on their shelves six weeks after they order. Your custom desk copies with their instructor supplements will arrive at your address at the same time.

To learn more about this new system for creating the perfect textbook, go to www.prenhall.com/esource. You can either go through the on-line walkthrough of how to create a book, or experiment yourself.

Supplements

Adopters of ESource receive an instructor's CD that contains professor and student code from the books in the series, as well as other instruction aides provided by authors. The website also holds approximately **350 Powerpoint transparencies** created by Jack Leifer of Univ. of Kentucky—Paducah available to download. Professors can either follow these transparencies as pre-prepared lectures or use them as the basis for their own custom presentations.

Titles in the ESource Series

Introduction to UNIX
0-13-095135-8
David I. Schwartz

Introduction to AutoCAD 2000
0-13-016732-0
Mark Dix and Paul Riley

Introduction to Maple
0-13-095133-1
David I. Schwartz

Introduction to Word
0-13-254764-3
David C. Kuncicky

Introduction to Excel, 2/e
0-13-016881-5
David C. Kuncicky

Introduction to Mathcad
0-13-937493-0
Ronald W. Larsen

Introduction to AutoCAD, R. 14
0-13-011001-9
Mark Dix and Paul Riley

Introduction to the Internet, 3/e
0-13-031355-6
Scott D. James

Design Concepts for Engineers
0-13-081369-9
Mark N. Horenstein

Engineering Design—A Day in the Life of Four Engineers
0-13-085089-6
Mark N. Horenstein

Engineering Ethics
0-13-784224-4
Charles B. Fleddermann

Engineering Success
0-13-080859-8
Peter Schiavone

Mathematics Review
0-13-011501-0
Peter Schiavone

Introduction to C
0-13-011854-0
Delores Etter

Introduction to C++
0-13-011855-9
Delores Etter

Introduction to MATLAB
0-13-013149-0
Delores Etter with David C. Kuncicky

www.prenhall.com/esource/   www.prenhall.com/esource/

Titles in the ESource Series

Introduction to FORTRAN 90
0-13-013146-6
Larry Nyhoff and Sanford Leestma

Introduction to Java
0-13-919416-9
Stephen J. Chapman

Introduction to Engineering Analysis
0-13-016733-9
Kirk D. Hagen

Introduction to PowerPoint
0-13-040214-1
Jack Leifer

Graphics Concepts
0-13-030687-8
Richard M. Lueptow

Graphics Concepts with Pro/ENGINEER
0-13-014154-2
Richard M. Lueptow, Jim Steger, and Michael T. Snyder

Graphics Concepts with SolidWorks
0-13-014155-0
Richard M. Lueptow and Michael Minbiole

Introduction to Visual Basic 6.0
0-13-026813-5
David I. Schneider

Introduction to Mathcad 2000
0-13-020007-7
Ronald W. Larsen

About the Authors

No project could ever come to pass without a group of authors who have the vision and the courage to turn a stack of blank paper into a book. The authors in this series worked diligently to produce their books, provide the building blocks of the series.

Delores M. Etter is a Professor of Electrical and Computer Engineering at the University of Colorado. Dr. Etter was a faculty member at the University of New Mexico and also a Visiting Professor at Stanford University. Dr. Etter was responsible for the Freshman Engineering Program at the University of New Mexico and is active in the Integrated Teaching Laboratory at the University of Colorado. She was elected a Fellow of the Institute of Electrical and Electronics Engineers for her contributions to education and for her technical leadership in digital signal processing. In addition to writing best-selling textbooks for engineering computing, Dr. Etter has also published research in the area of adaptive signal processing.

Sanford Leestma is a Professor of Mathematics and Computer Science at Calvin College, and received his Ph.D. from New Mexico State University. He has been the long-time co-author of successful textbooks on Fortran, Pascal, and data structures in Pascal. His current research interest are in the areas of algorithms and numerical computation.

Larry Nyhoff is a Professor of Mathematics and Computer Science at Calvin College. After doing bachelor's work at Calvin, and Master's work at Michigan, he received a Ph.D. from Michigan State and also did graduate work in computer science at Western Michigan. Dr. Nyhoff has taught at Calvin for the past 34 years—mathematics at first and computer science for the past several years. He has co-authored several computer science textbooks since 1981 including titles on Fortran and C++, as well as a brand new title on Data Structures in C++.

Acknowledgments: We express our sincere appreciation to all who helped in the preparation of this module, especially our acquisitions editor Alan Apt, managing editor Laura Steele, developmental editor Sandra Chavez, and production editor Judy Winthrop. We also thank Larry Genalo for several examples and exercises and Erin Fulp for the Internet address application in Chapter 10. We appreciate the insightful review provided by Bart Childs. We thank our families—Shar, Jeff, Dawn, Rebecca, Megan, Sara, Greg, Julie, Joshua, Derek, Tom, Joan; Marge, Michelle, Sandy, Lory, Michael—for being patient and understanding. We thank God for allowing us to write this text.

Mark Dix began working with AutoCAD in 1985 as a programmer for CAD Support Associates, Inc. He helped design a system for creating estimates and bills of material directly from AutoCAD drawing databases for use in the automated conveyor industry. This system became the basis for systems still widely in use today. In 1986 he began collaborating with Paul Riley to create AutoCAD training materials, combining Riley's background in industrial design and training with Dix's background in writing, curriculum development, and programming. Dix and Riley have created tutorial and teaching methods for every AutoCAD release since Version 2.5. Mr. Dix has a Master of Education from the University of Massachusetts. He is currently the Director of Dearborn Academy High School in Arlington, Massachusetts.

Paul Riley is an author, instructor, and designer specializing in graphics and design for multimedia. He is a founding partner of CAD Support Associates, a contract service and professional training organization for computer-aided design. His 15 years of business experience and 20 years of teaching experience are supported by degrees in education and computer science. Paul has taught AutoCAD at the University of Massachusetts at Lowell and is presently teaching AutoCAD at Mt. Ida College in Newton, Massachusetts. He has developed a program, Computer-aided Design for Professionals that is highly regarded by corporate clients and has been an ongoing success since 1982.

Scott D. James is a staff lecturer at Kettering University (formerly GMI Engineering & Management Institute) in Flint, Michigan. He is currently pursuing a Ph.D. in Systems Engineering with an emphasis on software engineering and computer-integrated manufacturing. Scott decided on writing textbooks after he found a void in the books that were available. "I really wanted a book that showed how to do things in good detail but in a clear and concise way. Many of the books on the market are full of fluff and force you to dig out the really important facts." Scott decided on teaching as a profession after several years in the computer industry. "I thought that it was really important to know what it was like outside of academia. I wanted to provide students with classes that were up to date and provide the information that is really used and needed."
Acknowledgments: Scott would like to acknowledge his family for the time to work on the text and his students and peers at Kettering who offered helpful critiques of the materials that eventually became the book.

Charles B. Fleddermann is a professor in the Department of Electrical and Computer Engineering at the University of New Mexico in Albuquerque, New Mexico. All of his degrees are in electrical engineering: his Bachelor's degree from the University of Notre Dame, and the Master's and Ph.D. from the University of Illinois at Urbana-Champaign. Prof. Fleddermann developed an engineering ethics course for his department in response to the ABET requirement to incorporate ethics topics into the undergraduate engineering curriculum. *Engineering Ethics* was written as a vehicle for presenting ethical theory, analysis, and problem solving to engineering undergraduates in a concise and readily accessible way.
Acknowledgments: I would like to thank Profs. Charles Harris and Michael Rabins of Texas A & M University whose NSF sponsored workshops on engineering ethics got me started thinking in this field. Special thanks to my wife Liz, who proofread the manuscript for this book, provided many useful suggestions, and who helped me learn how to teach "soft" topics to engineers.

David I. Schwartz is an Assistant Professor in the Computer Science Department at Cornell University and earned his B.S., M.S., and Ph.D. degrees in Civil Engineering from State University of New York at Buffalo. Throughout his graduate studies, Schwartz combined principles of computer science to applications of civil engineering. He became interested in helping students learn how to apply software tools for solving a variety of engineering problems. He teaches his students to learn incrementally and practice frequently to gain the maturity to tackle other subjects. In his spare time, Schwartz plays drums in a variety of bands.
Acknowledgments: I dedicate my books to my family, friends, and students who all helped in so many ways. Many thanks go to the schools of Civil Engineering and Engineering & Applied Science at State University of New York at Buffalo where I originally developed and tested my UNIX and Maple books. I greatly appreciate the opportunity to explore my goals and all the help from everyone at the Computer Science Department at Cornell. Eric Svendsen and everyone at Prentice Hall also deserve my gratitude for helping to make these books a reality. Many thanks, also, to those who submitted interviews and images.

Ron Larsen is an Associate Professor of Chemical Engineering at Montana State University, and received his Ph.D. from the Pennsylvania State University. He was initially attracted to engineering by the challenges the profession offers, but also appreciates that engineering is a serving profession. Some of the greatest challenges he has faced while teaching have involved non-traditional teaching methods, including evening courses for practicing engineers and teaching through an interpreter at the Mongolian National University. These experiences have provided tremendous opportunities to learn new ways to communicate technical material. He tries to incorporate the skills he has learned in non-traditional arenas to improve his lectures, written materials, and learning programs. Dr. Larsen views modern software as one of the new tools that will radically alter the way engineers work, and his book *Introduction to Mathcad* was written to help young engineers prepare to meet the challenges of an ever-changing workplace.
Acknowledgments: To my students at Montana State University who have endured the rough drafts and typos, and who still allow me to experiment with their classes—my sincere thanks.

Peter Schiavone is a professor and student advisor in the Department of Mechanical Engineering at the University of Alberta, Canada. He received his Ph.D. from the University of Strathclyde, U.K. in 1988. He has authored several books in the area of student academic success as well as numerous papers in international scientific research journals. Dr. Schiavone has worked in private industry in several different areas of engineering including aerospace and systems engineering. He founded the first Mathematics Resource Center at the University of Alberta, a unit designed specifically to teach new students the necessary *survival skills* in mathematics and the physical sciences required for success in first-year engineering. This led to the Students' Union Gold Key Award for outstanding contributions to the university. Dr. Schiavone lectures regularly to freshman engineering students and to new engineering professors on engineering success, in particular about maximizing students' academic performance. He wrote the book *Engineering Success* in order to share the *secrets of success in engineering study*: the most effective, tried and tested methods used by the most successful engineering students.

Acknowledgements: Thanks to Eric Svendsen for his encouragement and support; to Richard Felder for being such an inspiration; to my wife Linda for sharing my dreams and believing in me; and to Francesca and Antonio for putting up with Dad when working on the text.

Mark N. Horenstein is a Professor in the Department of Electrical and Computer Engineering at Boston University. He has degrees in Electrical Engineering from M.I.T. and U.C. Berkeley and has been involved in teaching engineering design for the greater part of his academic career. He  devised and developed the senior design project class taken by all electrical and computer engineering students at Boston University. In this class, the students work for a virtual engineering company developing products and systems for real-world engineering and social-service clients. Many of the design projects developed in his class have been aimed at assistive technologies for individuals with disabilities.

Acknowledgments: I would like to thank Prof. James Bethune, the architect of the Peak Performance event at Boston University, for his permission to highlight the competition in my text. Several of the ideas relating to brainstorming and teamwork were derived from a workshop on engineering design offered by Prof. Charles Lovas of Southern Methodist University. The principles of estimation were derived in part from a freshman engineering problem posed by Prof. Thomas Kincaid of Boston University.

Kirk D. Hagen is a professor at Weber State University in Ogden, Utah. He has taught introductory-level engineering courses and upper-division thermal science courses at WSU since 1993. He received his B.S. degree in physics from Weber State College and his M.S. degree  in mechanical engineering from Utah State University, after which he worked as a thermal designer/analyst in the aerospace and electronics industries. After several years of engineering practice, he resumed his formal education, earning his Ph.D. in mechanical engineering at the University of Utah. Hagen is the author of an undergraduate heat transfer text. Having drawn upon his industrial and teaching experience, he strongly believes that engineering students must develop effective analytical problem solving abilities. His book, *Introduction to Engineering Analysis*, was written to help beginning engineering students learn a systematic approach to engineering analysis.

Richard M. Lueptow is the Charles Deering McCormick Professor of Teaching Excellence and Associate Professor of Mechanical Engineering at Northwestern University. He is a native of Wisconsin and received his doctorate from the Massachusetts Institute of Technology in 1986. He teaches design, fluid mechanics, and  spectral analysis techniques. "In my design class I saw a need for a self-paced tutorial for my students to learn CAD software quickly and easily. I worked with several students a few years ago to develop just this type of tutorial, which has since evolved into a book. My goal is to introduce students to engineering graphics and CAD, while showing them how much fun it can be." Rich has an active research program on rotating filtration, Taylor Couette flow, granular flow, fire suppression, and acoustics. He has five patents and over 40 refereed journal and proceedings papers along with many other articles, abstracts, and presentations.

Acknowledgments: Thanks to my talented and hard-working co-authors as well as the many colleagues and students who took the tutorial for a "test drive." Special thanks to Mike Minbiole for his major contributions to Graphics Concepts with SolidWorks. Thanks also to Northwestern University for the time to work on a book. Most of all, thanks to my loving wife, Maiya, and my children, Hannah and Kyle, for supporting me in this endeavor. (Photo courtesy of Evanston Photographic Studios, Inc.)

Jack Leifer is an Assistant Professor in the Department of Mechanical Engineering at the University of Kentucky Extended Campus Program in Paducah, and was previously with the Department of Mathematical Sciences and Engineering at the University of South Carolina—  Aiken. He received his Ph.D. in Mechanical Engineering from the University of Texas at Austin in December 1995.

His current research interests include the modeling of sensors for manufacturing, and the use of Artificial Neural Networks to predict corrosion.

Acknowledgements: I'd like to thank my colleagues at USC—Aiken, especially Professors Mike May and Laurene Fausett, for their encouragement and feedback; Eric Svendsen and Joe Russo of Prentice Hall, for their useful suggestions and flexibility with deadlines; and my parents, Felice and Morton Leifer, for being there and providing support (as always) as I completed this book.

David C. Kuncicky is a native Floridian. He earned his Baccalaureate in psychology, Master's in computer science, and Ph.D. in computer science from Florida State University. He has served as a faculty member in the Department of Electrical Engineering at the FAMU–FSU College of Engineering and the Department of Computer Science at Florida State University. He has taught computer science and computer engineering courses for over 15 years. He has published research in the areas of intelligent hybrid systems and neural networks. He is currently the Director of Engineering at Bioreason, Inc. in Sante Fe, New Mexico.

Acknowledgments: Thanks to Steffie and Helen for putting up with my late nights and long weekends at the computer. Thanks also to the helpful and insightful technical reviews by Jerry Ralya, Kathy Kitto, Avi Singhal, Thomas Hill, Ron Eaglin, Larry Richards, and Susan Freeman. I appreciate the patience of Eric Svendsen and Joe Russo of Prentice Hall for gently guiding me through this project. Finally, thanks to Susan Bassett for having faith in my abilities, and for providing continued tutelage and support.

Jim Steger is currently Chief Technical Officer and cofounder of an Internet applications company. He graduated with a Bachelor of Science degree in Mechanical Engineering from Northwestern University. His prior work included mechanical engineering assignments at Motorola and Acco Brands. At Motorola, Jim worked on part design for two-way radios and was one of the lead mechanical engineers on a cellular phone product line. At Acco Brands, Jim was the sole engineer on numerous office product designs. His Worx stapler has won design awards in the United States and in Europe. Jim has been a Pro/Engineer user for over six years.

Acknowledgments: Many thanks to my co-authors, especially Rich Lueptow for his leadership on this project. I would also like to thank my family for their continuous support.

David I. Schneider holds an A.B. degree from Oberlin College and a Ph.D. degree in Mathematics from MIT. He has taught for 34 years, primarily at the University of Maryland. Dr. Schneider has authored 28 books, with one-half of them computer programming books. He has developed three customized software packages that are supplied as supplements to over 55 mathematics textbooks. His involvement with computers dates back to 1962, when he programmed a special purpose computer at MIT's Lincoln Laboratory to correct errors in a communications system.

Michael T. Snyder is President of Internet startup Appointments123.com. He is a native of Chicago, and he received his Bachelor of Science degree in Mechanical Engineering from the University of Notre Dame. Mike also graduated with honors from Northwestern University's Kellogg Graduate School of Management in 1999 with his Masters of Management degree. Before Appointments123.com, Mike was a mechanical engineer in new product development for Motorola Cellular and Acco Office Products. He has received four patents for his mechanical design work. "Pro/Engineer was an invaluable design tool for me, and I am glad to help students learn the basics of Pro/Engineer."

Acknowledgments: Thanks to Rich Lueptow and Jim Steger for inviting me to be a part of this great project. Of course, thanks to my wife Gretchen for her support in my various projects.

Stephen J. Chapman received a BS in Electrical Engineering from Louisiana State University (1975), an MSE in Electrical Engineering from the University of Central Florida (1979), and pursued further graduate studies at Rice University. Mr. Chapman is currently Manager of Technical Systems for British Aerospace Australia, in Melbourne, Australia. In this position, he provides technical direction and design authority for the work of younger engineers within the company. He is also continuing to teach at local universities on a part-time basis.

Mr. Chapman is a Senior Member of the Institute of Electrical and Electronics Engineers (and several of its component societies). He is also a member of the Association for Computing Machinery and the Institution of Engineers (Australia).

Reviewers

ESource benefited from a wealth of reviewers who on the series from its initial idea stage to its completion. Reviewers read manuscripts and contributed insightful comments that helped the authors write great books. We would like to thank everyone who helped us with this project.

Concept Document

Naeem Abdurrahman *University of Texas, Austin*
Grant Baker *University of Alaska, Anchorage*
Betty Barr *University of Houston*
William Beckwith *Clemson University*
Ramzi Bualuan *University of Notre Dame*
Dale Calkins *University of Washington*
Arthur Clausing *University of Illinois at Urbana–Champaign*
John Glover *University of Houston*
A.S. Hodel *Auburn University*
Denise Jackson *University of Tennessee, Knoxville*
Kathleen Kitto *Western Washington University*
Terry Kohutek *Texas A&M University*
Larry Richards *University of Virginia*
Avi Singhal *Arizona State University*
Joseph Wujek *University of California, Berkeley*
Mandochehr Zoghi *University of Dayton*

Books

Stephen Allan *Utah State University*
Naeem Abdurrahman *University of Texas, Austin*
Anil Bajaj *Purdue University*
Grant Baker *University of Alaska—Anchorage*
Betty Burr *University of Houston*
William Beckwith *Clemson University*
Haym Benaroya *Rutgers University*
Tom Bledsaw *ITT Technical Institute*
Tom Bryson *University of Missouri, Rolla*
Ramzi Bualuan *University of Notre Dame*
Dan Budny *Purdue University*
Dale Calkins *University of Washington*
Arthur Clausing *University of Illinois*
James Devine *University of South Florida*
Patrick Fitzhorn *Colorado State University*
Dale Elifrits *University of Missouri, Rolla*
Frank Gerlitz *Washtenaw College*
John Glover *University of Houston*
John Graham *University of North Carolina—Charlotte*
Malcom Heimer *Florida International University*
A.S. Hodel *Auburn University*
Vern Johnson *University of Arizona*
Kathleen Kitto *Western Washington University*
Robert Montgomery *Purdue University*
Mark Nagurka *Marquette University*
Romarathnam Narasimhan *University of Miami*
Larry Richards *University of Virginia*
Marc H. Richman *Brown University*
Avi Singhal *Arizona State University*
Tim Sykes *Houston Community College*
Thomas Hill *SUNY at Buffalo*
Michael S. Wells *Tennessee Tech University*
Joseph Wujek *University of California, Berkeley*
Edward Young *University of South Carolina*
Mandochehr Zoghi *University of Dayton*
John Biddle *California State Polytechnic University*
Fred Boadu *Duke University*
Harish Cherukuri *University of North Carolina–Charlotte*
Barry Crittendon *Virginia Polytechnic and State University*
Ron Eaglin *University of Central Florida*
Susan Freeman *Northeastern University*
Frank Gerlitz *Washtenaw Community College*
Otto Gygax *Oregon State University*
Donald Herling *Oregon State University*
James N. Jensen *SUNY at Buffalo*
Autar Kaw *University of South Florida*
Kenneth Klika *University of Akron*
Terry L. Kohutek *Texas A&M University*
Melvin J. Maron *University of Louisville*
Soronadi Nnaji *Florida A&M University*
Michael Peshkin *Northwestern University*
Randy Shih *Oregon Institute of Technology*
Neil R. Thompson *University of Waterloo*
Garry Young *Oklahoma State University*

Contents

ABOUT ESOURCE   v

ABOUT THE AUTHORS   ix

1   ENGINEERING GRAPHICS   1

1.1   The Importance of Engineering Graphics   1
1.2   Engineering Graphics   3
1.3   CAD   6
1.4   Design and CAD   7

2   PROJECTIONS USED IN ENGINEERING GRAPHICS   11

2.1   Projections   11
2.2   3-D Projections   12
2.3   Multiview Projections   14
2.4   Working Drawings   17

3   FREEHAND SKETCHING   23

3.1   Why Freehand Sketches?   23
3.2   Freehand Sketching Fundamentals   24
3.3   Basic Freehand Sketching   25
    3.3.1   Oblique Sketching   26
    3.3.2   Isometric Sketching   27
3.4   Advanced Freehand Sketching   28
    3.4.1   Freehand Oblique Sketching   28
    3.4.2   Isometric Sketching   30
    3.4.3   Orthographic Sketching   32

4   COMPUTER AIDED DESIGN AND DRAFTING   37

4.1   CAD Models   37
4.2   CAD and Solids Modeling   40
4.3   The Nature of Solids Modeling   42

5   STANDARD PRACTICE FOR ENGINEERING DRAWINGS   49

5.1   Introduction to Drawing Standards   49
5.2   Sheet Layouts   50
5.3   Lines   51
5.4   Dimension Placement and Conventions   52
5.5   Section and Detail Views   56
5.6   Fasteners and Screw Threads   59
5.7   Assembly Drawings   61

6 TOLERANCES 67

- 6.1 Why Tolerances? 67
- 6.2 Displaying Tolerances on Drawings 69
- 6.3 How to Determine Tolerances 70
- 6.4 Surface Finish 73
- 6.5 Geometric Dimensioning and Tolerancing 74

7 GETTING STARTED IN SOLIDWORKS® 85

- 7.1 Introduction and Reference 85
  - 7.1.1 Starting SolidWorks 87
  - 7.1.2 Checking the Options Settings 87
  - 7.1.3 Getting Help 90
- 7.2 Modeling the Guard 90
  - 7.2.1 Creating a New Part 90
  - 7.2.2 Sketching 91
  - 7.2.3 Dimensioning the Sketch 96
  - 7.2.4 Changing Values of the Dimensions 98
  - 7.2.5 Adding Fillets to a Sketch 100
  - 7.2.6 Extruding the Cross Section 100
  - 7.2.7 Viewing the Guard 102
  - 7.2.8 Cutting a Hole in the Guard 104
  - 7.2.9 Creating a Round at the Corners 106
- 7.3 Modeling the Arm 108
  - 7.3.1 Creating a New Part and a New Sketch 109
  - 7.3.2 Sketching the Lines 109
  - 7.3.3 Dimensioning the Sketch 110
  - 7.3.4 Adding a Relation 110
  - 7.3.5 Extruding the Arm 112
  - 7.3.6 Rounding the End 113
  - 7.3.7 Adding Chamfers to the End of the Arm 114
  - 7.3.8 Adding a Hole in the Arm 116
- 7.4 Modeling the Blade 118
  - 7.4.1 Sketching the Blade 118
  - 7.4.2 Extruding the Sketch 119
  - 7.4.3 Adding a Chamfer to Form the Edge of the Blade 120
  - 7.4.4 Mirroring a Feature 121
  - 7.4.5 Changing the Definition of the Extrusion 122

8 MODELING PARTS IN SOLIDWORKS: REVOLVES 125

- 8.1 Modeling the Cap 125
  - 8.1.1 Sketching the Cap 125
  - 8.1.2 Adding a Centerline 127
  - 8.1.3 Revolving the Sketch 128
- 8.2 Modeling the Handle 130
  - 8.2.1 Sketching the Base Feature of the Handle 130
  - 8.2.2 Revolving the Cross Section 132
  - 8.2.3 Sketching and Cutting a Single Groove 134
  - 8.2.4 Modeling More Grooves: Inserting a Linear Pattern 136
  - 8.2.5 Modeling the Rectangular Hole 138
  - 8.2.6 Modeling the Circular Hole 140
  - 8.2.7 Adding the Finishing Touches to the Handle 141
  - 8.2.8 Changing the Color of the Handle 143

8.3 Modeling the Rivet 144
    8.3.1 Creating Arcs 146
    8.3.2 Drawing the Rest of the Sketch 147
    8.3.3 Revolving the Sketch 148
    8.3.4 Reopening the Sketch and Adding Dimensions 149
    8.3.5 Changing the Detailing Settings 150
    8.3.6 Rounding the Bottom of the Rivet 152

9   Modeling an Assembly: The Pizza Cutter 155

9.1 Modeling the Cutter Sub-Assembly 155
    9.1.1 Creating a New Assembly Document 156
    9.1.2 Bringing the Rivet into the Sub-assembly 157
    9.1.3 Bringing the Arm into the Assembly and Orienting It 158
    9.1.4 Adding a Concentric Mate 160
    9.1.5 Adding a Coincident Mate 161
    9.1.6 Assembling the Blade 162
    9.1.7 Assembling the Second Arm 164
    9.1.8 Adding a Parallel Mate 164

9.2 Modeling the Pizza Cutter Assembly 166
    9.2.1 Creating a New Assembly and Inserting the Handle 166
    9.2.2 Assembling the Cap 166
    9.2.3 Assembling the Guard 168
    9.2.4 Hiding an Object 169
    9.2.5 Inserting the Sub-assembly into the Assembly 169
    9.2.6 Showing the Guard and Finishing the Assembly 172

9.3 Checking the Assembly for Interference 174
    9.3.1 Checking for Interference Volumes 174
    9.3.2 Opening a Sketch Within the Guard 175
    9.3.3 Creating the Rectangular Hole 176
    9.3.4 Adding Equations to Dimensions 177
    9.3.5 Rebuilding the Pizza Cutter Assembly with the New Guard 178

9.4 Creating an Exploded View of the Assembly 179
    9.4.1 Exploding the Cutter Sub-assembly from the Handle 180
    9.4.2 Exploding an Arm 182
    9.4.3 Exploding the Other Components 183

10   Creating Working Drawings 185

10.1 Detail Drawings in Solidworks 185
10.2 Editing a Drawing Sheet Format 186
    10.2.1 Creating a New Drawing Document 187
    10.2.2 Checking the Options Settings 188
    10.2.3 The Drawing Environment 190
    10.2.4 Modifying the Format's Text 191
    10.2.5 Modifying the Sheet Format's Lines 193

10.3 Creating a Drawing of the Arm 195
    10.3.1 Placing Orthographic Views 196
    10.3.2 Adding a Named View 197
    10.3.3 Adjusting the Views on the Sheet 198
    10.3.4 Adding Dimensions to the Orthographic Views 200
    10.3.5 Modifying a Dimension's Text 202
    10.3.6 Specifying Tolerances 203
    10.3.7 Changing Dimensions in the Drawing and the Part 204

10.4	Creating a Drawing of The Pizza Cutter Assembly 206
	10.4.1 Setting Up the Pizza Cutter Drawing 207
	10.4.2 Adding Orthographic and Isometric Views to the Drawing 207
	10.4.3 Adding a Section View 209
	10.4.4 Adding a Detail View 210
	10.4.5 Adding Numbers to the Components of the Assembly 211
	10.4.6 Adding a Bill of Materials 213

INDEX 217

1
Engineering Graphics

Overview

Engineering designs start as images in the mind's eye of an engineer. Engineering graphics has evolved to communicate and record these ideas on paper both two- and three-dimensionally. In the past few decades, the computer has made it possible to automate the creation of engineering graphics. Today engineering design and engineering graphics are inextricably connected. Engineering design is communicated visually using engineering graphics.

1.1 THE IMPORTANCE OF ENGINEERING GRAPHICS

"Visualizing" a picture or image in your mind is a familiar experience. The image can be visualized at many different levels of abstraction. Think about light and you might see the image of a light bulb in your "mind's eye." Alternatively, you might think about light versus dark. Or you might visualize a flashlight or table lamp. Such visual thinking is necessary in engineering and science. Albert Einstein said that he rarely thought in words. Instead, he laboriously translated his visual images into verbal and mathematical terms.

Visual thinking is a foundation of engineering. Walter P. Chrysler, founder of the automobile company, recounted his experience as an apprentice machinist where he built a model locomotive that existed "within my mind so real, so complete, that it seemed to have three dimensions there."

Sections
- 1.1 The Importance of Engineering Graphics
- 1.2 Engineering Graphics
- 1.3 CAD
- 1.4 Design and CAD

Objectives

After reading this chapter, you should be able to:

- Describe visual thinking
- Differentiate perspective, isometric, and orthographic projections
- Understand the basis of CAD
- Understand the relationship between design and CAD

Yet, the complexity of today's technology rarely permits a single person to build a device from his own visual image. The images must be conveyed to other engineers and designers. In addition, those images must be constructed in such a way that they are in a readily recognizable, consistent, and readable format. This assures that the visual ideas are clearly and unambiguously conveyed to others. *Engineering graphics* is a highly stylized way of presenting images of parts or assemblies.

A major portion of engineering information is recorded and transmitted using engineering graphics. In fact, 92 percent of the design process is graphically based. Written and verbal communications along with mathematics account for the remaining eight percent. To demonstrate the effectiveness of engineering graphics compared to a written description try to visualize an ice scraper based on this word description:

> An ice scraper is generally in the shape of a 140 × 80 × 10 mm rectangular prism. One end is beveled from zero thickness to the maximum thickness in a length of 40 mm to form a sharp edge. The opposite end is semicircular. A 20 mm diameter hole is positioned so the center of the hole is 40 mm from the semicircular end and 40 mm from either side of the scraper.

It is evident immediately that the shape of the ice scraper is much more easily visualized from the graphical representation shown in Figure 1.1 than from the word description. Humans grasp information much more quickly when that information is presented in a graphical or visual form rather than as a word description.

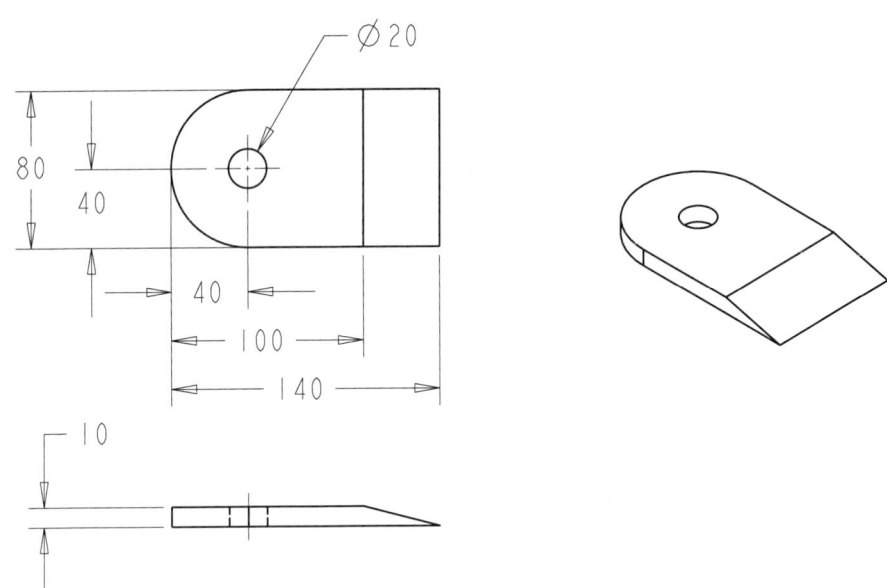

Figure 1.1.

Engineering drawings, whether done using a pencil and paper or a computer, start with a blank page or screen. The engineer's mind's eye image must be transferred to the paper or computer screen. The creative nature of this activity is similar to that of an artist. Perhaps the greatest example of this is Leonardo da Vinci, who had exceptional engineering creativity devising items such as parachutes and ball bearings, shown in Figure 1.2, hundreds of years before they were re-invented. He also had exceptional artistic talent, creating some of the most famous pictures ever painted such as *Mona Lisa* and *The Last Supper*.

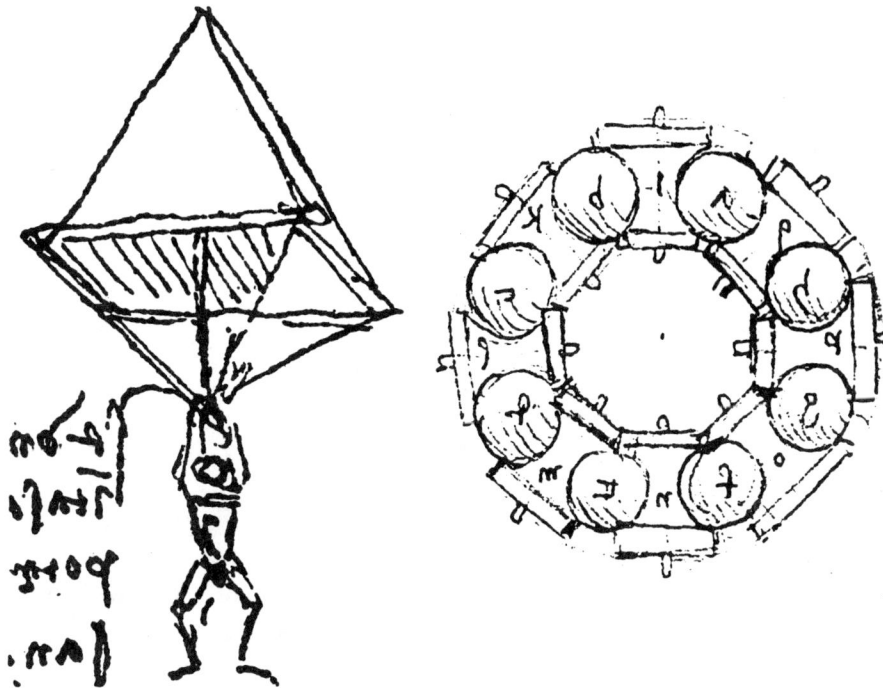

Figure 1.2. The creative genius of da Vinci is evident in these sketches of a parachute and a ball bearing, both devised hundreds of years before they were re-invented. (Parachute used with permission of the Biblioteca–Pinacoteca Ambrosiana, Milan, Italy. Property of the Ambrosian Library. All rights reserved. Reproduction is forbidden. [*The Inventions of Leonardo da Vinci,* Charles Gibbs-Smith, Phaidon Press, Oxford, 1978, p. 24.] Ball bearing used with permission of EMB-Service for Publishers, Lucerne, Switzerland. [*The Unknown Leonardo,* edited by Ladislao Reti, McGraw-Hill, 1974, p. 286])

1.2 ENGINEERING GRAPHICS

The first authentic record of engineering graphics dates back to 2130 BC, based on a statue now in the Museum of the Louvre, Paris. The statue depicts an engineer and governor of a small city-state in an area later known as Babylon. At the base of the statue are measuring scales and scribing instruments along with a plan of a fortress engraved on a stone tablet.

Except for the use of pen and paper rather than stone tablets, it was not until printed books appeared around 1450 that techniques of graphics advanced. Around the same time, *pictorial perspective* drawing was invented by artist Paolo Uccello. This type of drawing presents an object much like it would look to the human eye or in a photograph, as shown in Figure 1.3. The essential characteristic of a perspective drawing is that parallel lines converge at a point in the distance like parallel railroad tracks seem to converge in the distance. Copper-plate engravings permitted the production of finely detailed technical drawings using pictorial perspective in large numbers. The pictorial perspective drawings were crucial to the advancement of technology through the Renaissance and until the beginning of the Industrial Revolution. But these drawings could not convey adequately details of the construction of an object. One solution to this problem was the use of the *exploded view* developed in the 15th century and perfected by Leonardo da Vinci. The exploded view of an assembly of individual parts shows the parts spread out along a common axis, as shown for the hoist in Figure 1.4. The exploded view reveals details of the individual parts along with showing the order in which they are assembled.

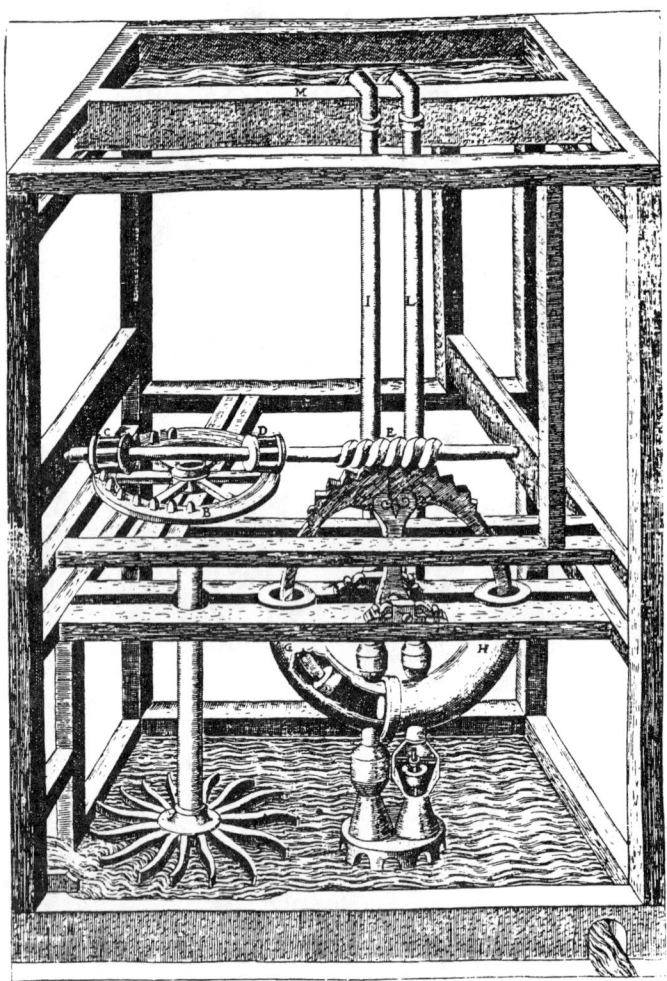

Figure 1.3. An example of pictorial perspective by Agostino Ramelli in 1588. (Used with permission of the Syndics of Cambridge University Library. [*The Various and Ingenious Machines of Agostino Ramelli (1588)*, translated by Martha Teach Gnudi, Johns Hopkins University Press, 1976. p. 83.])

The Industrial Revolution brought with it the need to tie more closely the concept of a design with the final manufactured product using technical drawing. The perspective drawing of a simple object in Figure 1.5a shows pictorially what the object looks like. However, it is difficult to represent accurately dimensions and other details in a perspective drawing. *Orthographic projections*, developed in 1528 by German artist Albrecht Dürer, accomplish this quite well. An orthographic projection typically shows three views of an object. Each view shows a different side of the object (say the front, top, and side). An example of an orthographic projection is shown in Figure 1.5b. Orthographic projections are typically easy to draw, and the lengths and angles in orthographic projections have little distortion. As a result, orthographic drawings can convey more information than a perspective drawing. But their interpretation takes more effort than a pictorial perspective, as is evident from Figure 1.5. French philosopher and mathematician René Descartes laid the foundation for the mathematical principles of projections by connecting geometry to algebra in the 17th century. Much later Gaspard Monge, a French mathematician, "invented" the mathematical principles

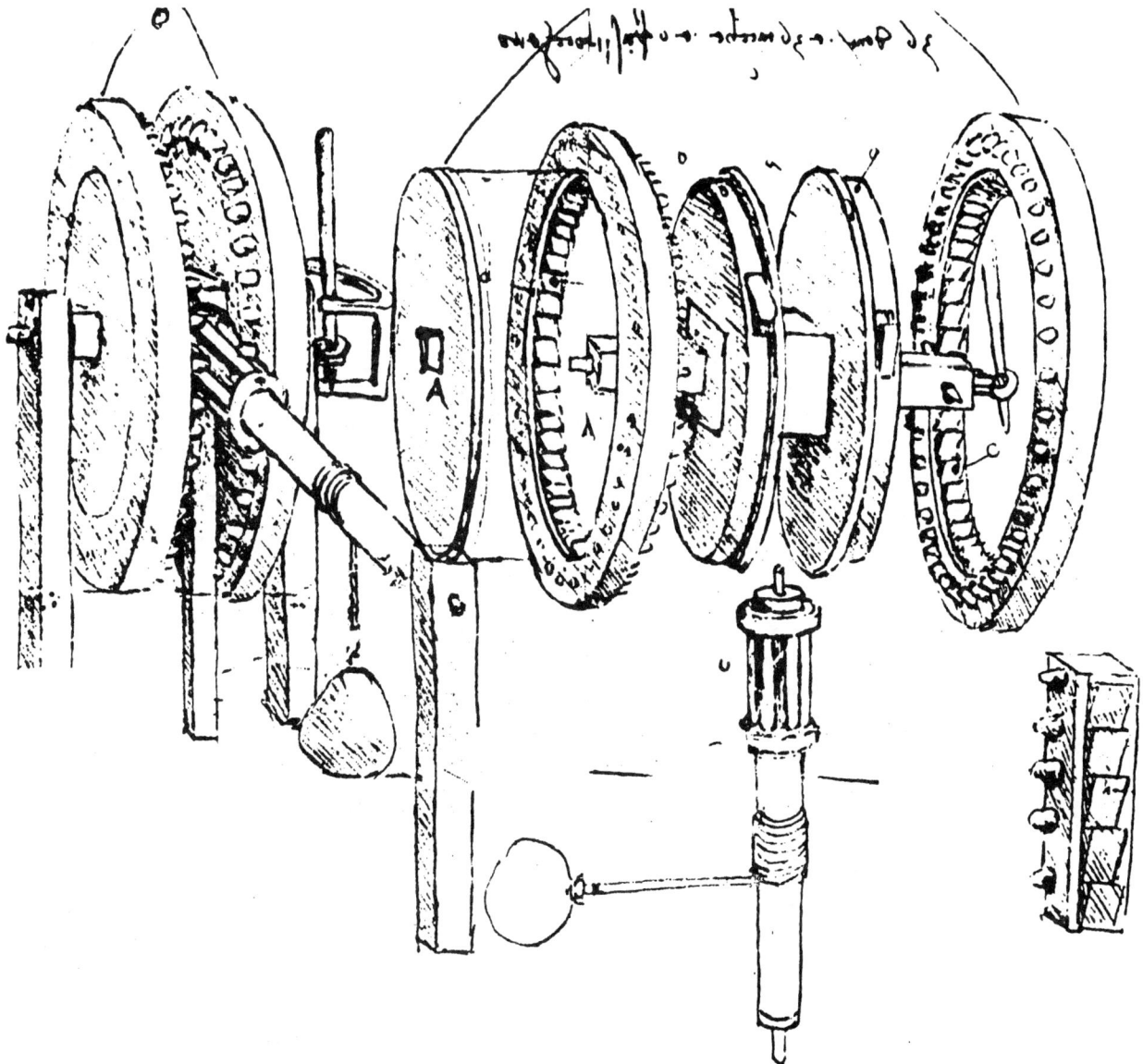

Figure 1.4. Assembled and exploded view of a hoist by Leonardo da Vinci, circa 1500. (Used with permission of the Biblioteca-Pinacoteca Ambrosiana, Milan, Italy. Property of the Ambrosian Library. All rights reserved. Reproduction is forbidden. [*The Inventions of Leonardo da Vinci*, Charles Gibbs-Smith, Phaidon Press, Oxford, 1978, p. 64.])

of projection known as *descriptive geometry*. These principles form the basis of engineering graphics today. But because these principles were thought to be of such strategic importance, they remained military secrets until 1795. By the 19th century, orthographic projections were used almost universally in mechanical drawing, and they are still the basis for engineering drawings today.

The *isometric view* is used more often today than the pictorial perspective view. Historically, it was used for centuries by engravers. Then in the early 19th century, William Farish, an English mathematician, formalized the isometric view and introduced it to engineers. The isometric view simplifies the pictorial perspective. In an isometric view, parallel lines remain parallel rather than converging to a point in the distance, as shown in Fig. 1.5c. Keeping parallel lines parallel distorts the appearance of the object slightly. But the distortion in an isometric view is negligible for objects of limited depth. For situations in which the depth of the object is large, such as an architectural view down a long hallway, the pictorial perspective is preferable. The advantage of the isometric view, though, is that it is much easier to draw than a pictorial perspective view.

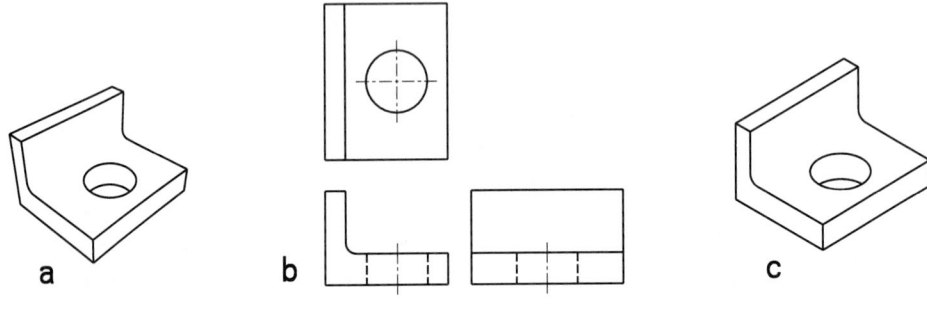

Figure 1.5.

1.3 CAD

The introduction of the computer revolutionized engineering graphics. Pioneers in computer-aided engineering graphics envisioned the computer as a tool to replace paper and pencil drafting with a system that is more automated, efficient, and accurate. The first demonstration of a computer-based drafting tool was a system called SKETCHPAD developed at the Massachusetts Institute of Technology in 1963 by Ivan Sutherland. The system used a monochrome monitor with a light pen for input from the user. The following year IBM commercialized computer-aided drafting.

During the 1970s, computer-aided drafting blossomed as the technology changed from scientific endeavor to an economically indispensable industrial tool for design. Commands for geometry generators to create commonly occurring shapes were added. Functions were added to control the viewing of the drawing geometry. Modifiers such as rotate, delete, and mirror were implemented. Commands could be accessed by typing on the keyboard or by using a mouse. Perhaps most importantly, three-dimensional modeling techniques became a key part of engineering graphics software.

By the 1980s, computer-aided drafting became fully developed in the marketplace as a standard tool in industry. In addition, the current technology of solids modeling came about. Solids models represent objects in the virtual environment of the computer just as they exist in reality, having a volume as well as surfaces and edges. The introduction of Pro/ENGINEER® in 1988 and SolidWorks® in the 1990s revolutionized computer-aided design and drafting. Today solids modeling remains the state-of-the-art technology.

What we have been referring to as computer-aided drafting is usually termed *CAD*, an acronym for Computer Aided Design, Computer Aided Drafting, or Computer Aided Design and Drafting. Originally the term Computer Aided Design included any technique that uses computers in the design process including drafting, stress analysis, and motion analysis. But over the last 35 years CAD has come to refer more specifically to Computer Aided Design and Drafting. Computer Aided Engineering (CAE) is used to refer to the broader range of computer-related design tools.

1.4 DESIGN AND CAD

Inextricably connected with engineering graphics is the *design process* in which an engineering or design team faces a particular engineering problem and devises a solution to that problem. Often the design team includes persons responsible for engineering, product design, production system design, manufacturing, marketing, and sales. The idea is to simultaneously develop the product and the manufacturing process for the product. This is known as *concurrent engineering* or *integrated product and process design*. Key to the success of concurrent engineering is communication of information. Engineering graphics is one of the primary methods used by concurrent engineering teams to record and transfer information during the design process.

The process of bringing a product to market is shown in Figure 1.6. The process begins with the identification of a market, or user need. After this, the design process is the key to bridging the gap between a user need and the manufacture and sales of a product. The design process can be broken down into three parts as indicated in Figure 1.6.

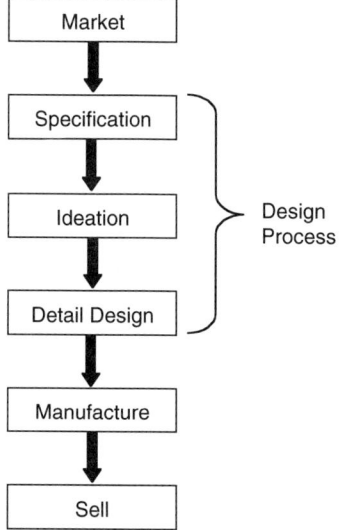

Figure 1.6.

The first part is the *specification* of the problem. The *design specification* is a list of requirements that the final product must meet, including size, performance, weight, and so on. The second part of the design process is called *ideation* or conceptual design. In this phase, the design team devises as many ideas for solutions to the design problem as possible and then narrows them down to the best one based on the specification. In many cases, a designer or engineer will quickly sketch ideas to explore or communicate the design concept to the rest of the design team. These freehand sketches form the basis for the details of the design that are laid down in the third phase of the design process. It is in this last phase of the *design process*, known as *detail design*, that CAD is crucial. The conceptual ideas for the product that were seen in a designer's mind's eye or are on paper as rough sketches must then be translated into the visual language of engineering graphics. In this way, the ideas can be understood clearly and accurately by the design team and other designers, engineers, fabricators, suppliers, and machinists. It is in this phase of the design that the nitty gritty details have to be worked out. Should the device be 30 mm long or should it be 35 mm long? How will one part fit with another? What size hole should be used? What material should be used? What manufacturing process will be used to make the part? The number of individual decisions that need to be made can be very large, even for a relatively simple part. Once the engineering drawings have been created, the product can be manufactured and eventually sold.

The use of CAD has had a great impact on the design process. For example, a part may be modified several times to meet the design specification or to mate with another part. Before the advent of CAD, these modifications were very tedious, time-consuming, and prone to error. However, CAD has made it possible to make these changes relatively easily and quickly. The connectivity of computers using local area networks then makes the revised electronic drawings available to a team of engineers at an instant. This is crucial as engineering systems become more complex and operational requirements become more stringent. For example, a modern jet aircraft has several million individual parts that must all fit together and perform safely for several decades.

Although CAD had a great impact on making the design process speedier and more accurate, the capabilities of the first few generations of CAD were still limited. Early CAD systems only provided a means of automating the drafting process to create orthographic engineering drawings. The designer or engineer would simply generate a line on the computer screen rather than drawing the line on paper. Current computer graphics software such as "paint" or "draw" programs for personal computers work this way. As CAD became more sophisticated, it helped automate the drafting process based on the "intelligence" of the software. CAD software lacking such intelligence required an engineer to draw a pair of parallel lines an exact distance apart by specifying coordinates of the endpoints of the lines. More advanced generations of CAD software permitted an engineer to draw approximately parallel lines using a mouse. Then the engineer would specify a particular distance between the lines and that the lines should be parallel. The CAD software then automatically placed these lines the specified distance apart and made them parallel. However, the major problem with the early generations of CAD software was that the designer or engineer was simply creating two-dimensional orthographic views of a three-dimensional part using a computer instead of a pencil and paper. From these two-dimensional views, the engineer still needed to reconstruct the mind's eye view of the three-dimensional image in the same way as if the drawings were created by hand.

The current generation of CAD software has had a very profound effect on the design process, because it is now possible to create a virtual prototype of a part or assembly on a computer. For example, the solids model of a pizza cutter is shown in Figure 1.7. Rather than translating a three-dimensional image from the mind's eye to a two-dimensional orthographic projection of the object, current CAD software starts with generating a three-dimensional virtual model of the object directly on the computer. This virtual model can be rotated so that it can be viewed from different angles. Several parts can be virtually assembled on the computer to make sure that they fit together. The assembled parts can be viewed as an assembly or in an exploded view. All of this is done in a virtual environment on the computer before the two-dimensional orthographic engineering drawings are even produced. It is still usually necessary to produce the orthographic engineering drawings. But these drawings only to serve as a standard means of engineering graphics communication, rather than as a tedious, time-consuming task necessary to proceed with the design process.

Figure 1.7.

PROFESSIONAL SUCCESS

What Happened to Pencil and Paper Drawings?

CAD drawings have already replaced pencil and paper drawings. Through the 1970s and even into the 1980s many engineering and design facilities consisted of rows and rows of drafting tables with a designer or engineer hunched over a drawing on each table. Engineering colleges and universities required a full-year course in "engineering drafting" or "graphics communication" for all engineering students. Many of these students purchased a set of drawing instruments along with their first semester textbooks. They spent endless hours practicing lettering and drawing perfect circles.

Now nearly all of the drafting tables and drafting courses have been replaced by CAD. One can still find a drafting table here and there, but it is not for creating engineering drawings. Most often it is used for displaying a large CAD drawing to designers and engineers. They can make notes on the drawing or freehand sketch modifications on the drawing. The pencil is still an ideal means for generating ideas and quickly conveying those ideas to others. But eventually, all of the pencil markings are used as the basis for modifying the CAD drawing.

In some companies with traditional products that have not changed for decades (like a spoon or a chair), pencil and paper drawings are gradually being converted into electronic form. In some cases, the original drawing is simply scanned to create an electronic version. The scanned drawing cannot be modified, but the electronic version takes much less storage space than a hard copy. In other cases, the original pencil and paper drawings are being systematically converted to CAD drawings, so that they can be modified if necessary. In any case, though, the pencil and paper drawing is now just a part of the history of engineering.

KEY TERMS

CAD
Descriptive geometry
Design process
Design specification

Detail Design
Engineering graphics
Exploded view
Ideation

Isometric view
Orthographic projection
Pictorial perspective view

Problems

1. Research and report on an important historical figure in engineering graphics such as Leonardo da Vinci, Albrecht Dürer, Gaspard Monge, René Descartes, William Farish, M. C. Escher, Frank Lloyd Wright, Ivan Sutherland, or Paolo Uccello.
2. Using the World Wide Web, search for computer-aided design web sites. Report on the variety of software available for CAD and its capabilities.
3. Trace the early development of CAD and report on your findings. A particularly useful starting point is an article by S. H. Clauser in the November 1981 issue of *Mechanical Engineering*.
4. Computer hardware and user interface development has had a profound effect on the evolution of CAD software. Research and report on the impact of the light pen, the graphics tablet, the direct view storage tube, raster graphics technology, the work-station computer, the personal computer, or the mouse on the capability and evolution of CAD. (Many textbooks on engineering graphics discuss these items.)
5. Outline the design process and indicate how the three steps in the design process would relate to:

    a. the design of a suspension system for a mountain bike.
    b. the design of an infant car seat.
    c. the design of headphones for a portable cassette player.
    d. the design of a squirrel-proof bird feeder.
    e. the design of a car-mounted bicycle carrier.
    f. the design of children's playground equipment.

6. Explain how "paint" or "draw" programs such as McDraw, Adobe Illustrator, MS Paint, or drawing tools in MS Word differ from CAD engineering graphics software.

2

Projections Used in Engineering Graphics

OVERVIEW

Engineering graphics is a highly stylized scheme to represent three-dimensional objects on a two-dimensional paper or computer screen. This can be accomplished by representing all three dimensions of an object in a single image or by presenting a collection of views of different sides of the object. Working drawings are the practical result of using engineering graphics to represent objects.

2.1 PROJECTIONS

The goal in engineering graphics, whether it is freehand sketching or CAD, is to represent a physical object or the mind's eye image of an object so that the image can be conveyed to other persons. Objects can be shown as *3-D projections* or *multiview projections*. Figure 2.1 shows the handle of a pizza cutter shown in both ways. The 3-D projection clearly suggests the three-dimensional character of the handle, even though it is displayed on a two-dimensional medium (the page). 3-D projections are useful in that they provide an image that is similar to the image in the designer's mind's eye. But 3-D projections are often weak in providing adequate details of the object, and there is often some distortion of the object. For instance, a circular hole at the right end of the handle becomes an ellipse in an isometric 3-D projection.

Multiview projections are used to overcome the weaknesses of 3-D projections. Multiview projections are a collection of flat 2-D drawings of the different sides of an object.

SECTIONS

- 2.1 Projections
- 2.2 3-D Projections
- 2.3 Multiview Projections
- 2.4 Working Drawings

OBJECTIVES

After reading this chapter, you should:

- Understand the spatial relation between 3-D projections and multiview projections
- Be able to differentiate isometric, trimetric, perspective, and oblique 3-D projections
- Understand how multiview projections are related to the views of the sides of an object
- Know the proper placement of orthographic views
- Know the difference between third-angle and first-angle orthographic projections
- Be able to differentiate the various kinds of working drawings

For instance, the side and bottom end of the pizza cutter handle are shown in the multi-view projection in Figure 2.1. Because there are two views, it is quite easy to depict details of the object. In addition, taken together, multiview projections provide a more accurate representation of the object than the 3-D projection—a circular hole appears as a circle in a multiview projection. On the other hand, multiview projections require substantial interpretation, and the overall shape of an object is often not obvious upon first glance. Consequently, the combination of the overall image provided by 3-D projections and details provided by multiview projections yield a representation of an object that is best. The shape of the object is immediately evident from the 3-D projection, and the detail needed for an accurate description of the object is available from the multiview projection.

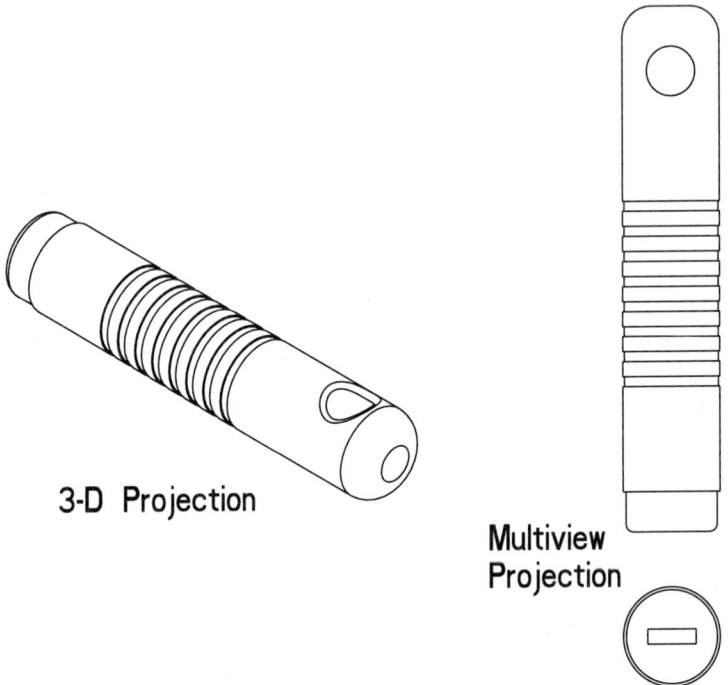

Figure 2.1.

2.2   3-D PROJECTIONS

Three different types of 3-D projections are available in most CAD software: isometric, trimetric, and perspective. These three views of a cube are shown in Figure 2.2. In all three cases, these 3-D projections represent all three dimensions of the cube in a single planar image. Although it is clear in all three cases that the object is a cube, each type of 3-D projection has its advantages and disadvantages.

The *isometric* projection has a standard orientation that makes it the typical projection used in CAD. In an isometric projection, the width and depth dimensions are sketched at 30° above horizontal as shown in Figure 2.2a. This results in the three angles at the upper front corner of the cube being equal to 120°. The three sides of the cube are also equal, leading to the term iso (equal) -metric (measure). Isometric drawings work quite well for objects of limited depth. However, an isometric drawing distorts the object when the depth is significant. In this case, a pictorial perspective drawing is better.

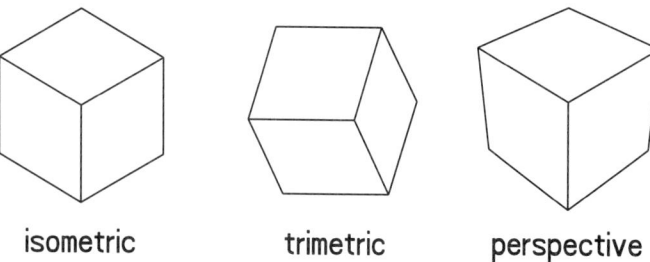

Figure 2.2.

In general, the *trimetric* projection offers more flexibility in orienting the object in space. The width and depth dimensions are at arbitrary angles to the horizontal, and the three angles at the upper front corner of the cube are unequal. This makes the three sides of the cube each have a different length as measured in the plane of the drawing; hence the name tri-metric. In most CAD software, the trimetric projection fixes one side along a horizontal line and tips the cube forward as shown in Figure 2.2b. A *dimetric* projection sets two sides of the cube, usually those of the front face, equal.

A pictorial perspective, or simply *perspective*, projection is drawn so that parallel lines converge in the distance as shown in Figure 2.2c, unlike isometric or trimetric projections where parallel lines remain parallel. A perspective projection is quite useful in providing a realistic image of an object when the object spans a long distance, such as the view of a bridge or aircraft from one end. Generally, small manufactured objects are adequately represented by isometric or trimetric views.

Two types of pictorial sketches are used frequently in freehand sketching: isometric and oblique. The isometric projection was discussed with respect to 3-D CAD projections. The isometric projection is often used in freehand sketching because it is relatively easy to create a realistic sketch of an object. But the *oblique projection* is usually even easier to sketch. The oblique projection places the principal face of the object parallel to the plane of the paper with the axes in the plane of the paper perpendicular to one another. The axis into the paper is at an arbitrary angle with respect to the horizontal. Figure 2.3 compares an isometric projection and an oblique projection of a cube with a hole in it. The advantage of the oblique projection is that details in the front face of the object retain their true shape. For instance, the circle on the front face is circular in the oblique projection, while it is elliptical in the isometric projection. This feature often makes oblique freehand sketching somewhat easier than isometric sketching.

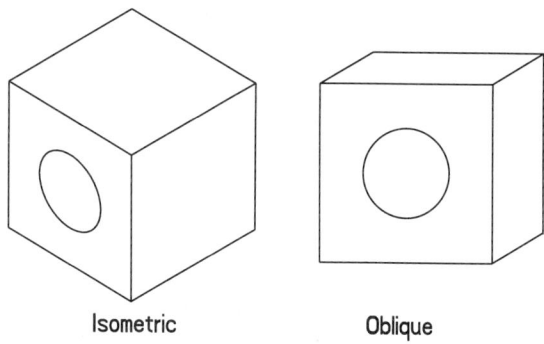

Figure 2.3.

2.3 MULTIVIEW PROJECTIONS

The standard means of multiview projection in engineering graphics is what we have referred to earlier as the *orthographic projection*. Although 3-D projections provide a readily identifiable visual image of an object, multiview projections are ideal for showing the details of an object. Dimensions can be shown easily and most features remain undistorted in multiview projections.

An orthographic projection is most easily thought of as a collection of views of different sides of an object—front, top, side, and so forth. For instance, two orthographic projections could be used for a coffee mug. The front view would show the sidewall of the mug along with the loop forming the handle. The top view would show what one would see looking down into the mug—a circular rim of the mug, the bottom of the inside of the mug, and the top of the handle that sticks out of the side of the mug. Dimensions of the mug could easily be added to the projections of each side of the mug to create an engineering drawing.

One useful way of looking at multiview projections is to imagine a glass box surrounding the object as shown in Figure 2.4. The image of each side of the object can be projected onto the wall of the glass box. Now an observer on the outside of the box can

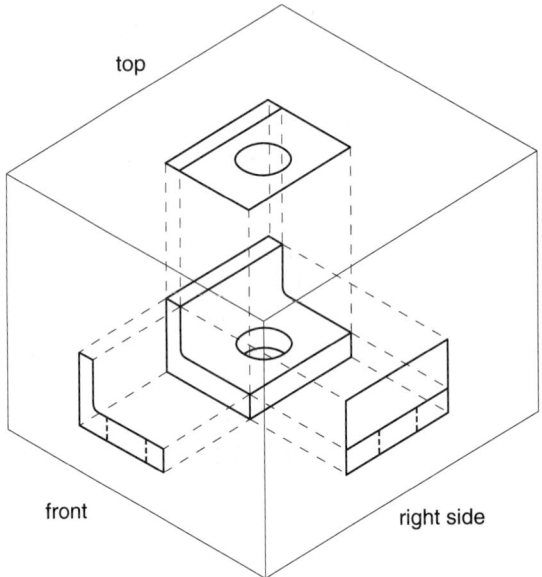

Figure 2.4.

see each side of the object as projected on each of the six walls of the box. Solid lines show the edges evident in the projection, and dashed lines show lines that are hidden by the object. Now imagine unfolding the glass box as if each of the edges of the glass box were a hinge, so that the front view is in the middle. Now the unfolded glass box represents all six sides of the object in a single plane as shown in Figure 2.5. In unfolding the glass box, the top view is positioned above the front view, the bottom view is below the front view, the right-side view is to the right of the front view, and so on. The dimensions of the object remain the same in all views. For example, the horizontal dimension (width of the object) in the front view is identical in that same dimension in the top view and bottom view. The views also remain aligned so that the bottom edge in the front

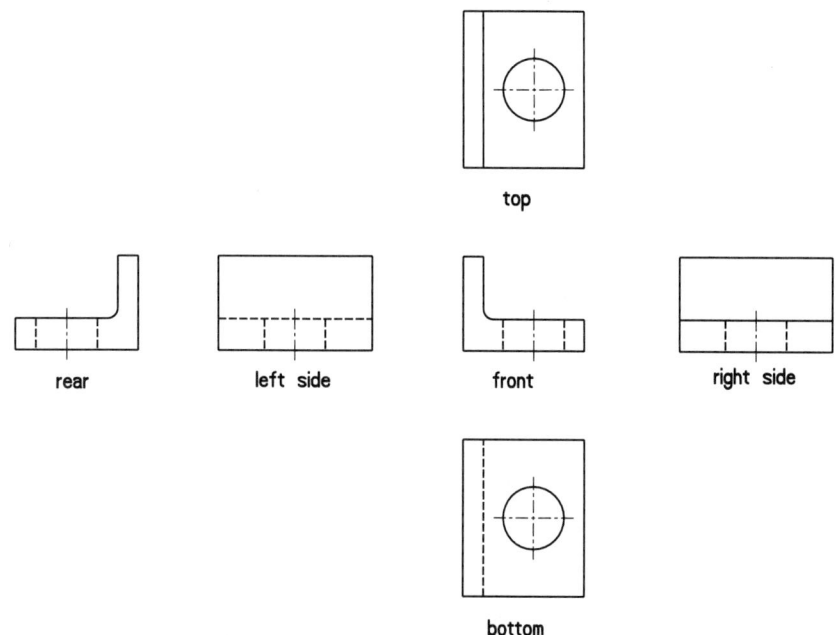

Figure 2.5.

view is even with the bottom edge in the right side, left side, and rear views. Likewise, the top edges remain aligned. Finally, the same edges in adjacent views are closest together. For instance, the same edge of the object is at the left-side of the front view and the right side of the left side view. This edge in the front view is closest to the same edge in the left-side view.

In many cases, three views are needed to represent an object accurately, although in some cases (like a coffee mug) only two views are necessary, and in other cases more than three views are needed to show complex features of the object. It is helpful to select the side of the object that is most descriptive of the object as the front view. Sometimes this may place an object so that what is normally thought of as the front of the object is not shown in the front view of the multiview projection. For example, what is usually described as being the side of a car should be chosen as the front view, because this view is probably most descriptive and easily recognizable as a car. A view of the front of a car (grille, bumper, and windshield) is not as descriptive or as obvious as the side of the car. Furthermore, the object should be properly oriented in the front view. For instance, a car should be shown with its wheels downward in their normal operating position for the front view. The other views that are shown in addition to the front view should be views that best represent features of the object. Normally the minimum number of views necessary to accurately represent the object is used. The standard practice is to use the front, top, and right-side views. But the choice of which views to use depends on the object and which details need to be shown most clearly.

A complication that arises in multiview projections is that two different standards are used for the placement of projections. In North America (and to some extent, in Great Britain) the unfolding of the glass box approach places the top view above the front view, the right-side view to the right of the front view, and so on. This placement of views is called *third-angle projection*. But in most of the rest of the world an alternative approach for the placement of views is used. In this case the placement of views is what would result if the object were laid on the paper with its front side up for the front view and then rolled on one edge for the other views. For instance, if the object were rolled to the right so that it rests on its right side, then the left side would be facing up. So the left-side view is placed to the right of the front view. Likewise, if the object were lying on the paper with the front view up and then rolled toward the bottom of the paper, it would be resting on its bottom side, so that the top side faces upward. Thus, the top view is placed below the front view. This placement of views, known as the *first-angle projection*, simply reverses the location of the top and bottom views and the location of the left-side and right-side views with respect to the front view compared to the third-angle projection. The views themselves remain the same in both projections.

Although, the difference between the two projections is only in the placement of the views, great potential for confusion and manufacturing errors can result in engineering drawings that are used globally. To avoid misunderstanding, international projection symbols, shown in Figure 2.6, have been developed to distinguish between third-angle and first-angle projections on drawings. The symbol shows two views of a truncated cone. In the first-angle projection symbol, the truncated end of the cone (two concentric circles) is placed on the base side of the cone, as it would be in a first-angle projection. In the third-angle projection symbol, the truncated end of the cone is placed on the truncated side of the cone, as it would be in a third-angle projection. Usually these symbols appear in or near the title block of the drawing when the possibility of confusion is anticipated. Most CAD software automatically uses the third-angle projection for engineering drawings.

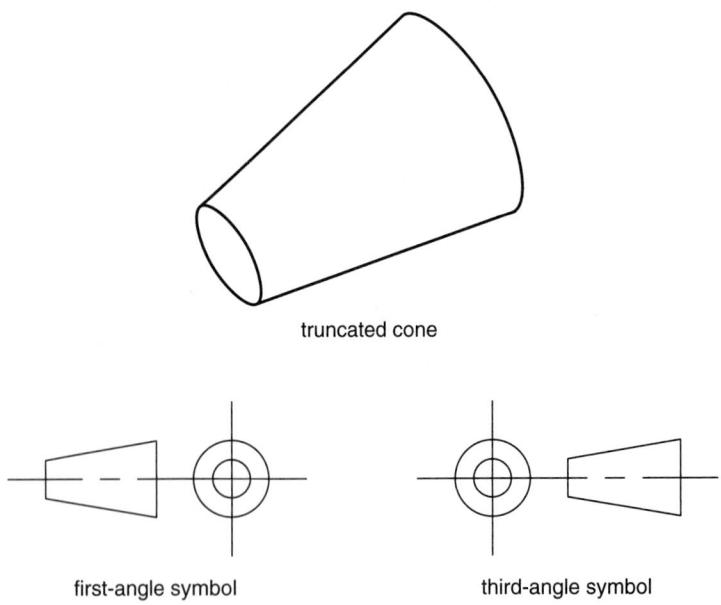

Figure 2.6.

A problem that frequently occurs in orthographic projections is that one of the faces of the object is at an angle to the orthographic planes that form the imaginary glass box. An example is the object shown in Figure 2.7. The circular hole with a keyway (rectangular cutout) that is perpendicular to the angled face appears in both the top view and the right-side view. However, it is distorted in both of these views, because it is on an angled plane of the object. An auxiliary view is used to avoid this distortion. In this case, a view of the object is drawn so that the angled face is parallel to the auxiliary view plane. The view is based on the viewer looking at the object along a line of sight that is perpendicular to the angled face. When viewed in this direction, the circular hole in the auxiliary view appears as a circle without any distortion. As suggested by the dash-dot lines in Figure 2.7, the auxiliary view is projected from the front view in the same way as the top and right-side views are projected. Thus, its position with respect to the front view depends on the orientation of the angled face. It is normal practice not to project hidden lines or other features that are not directly related to the angled surface.

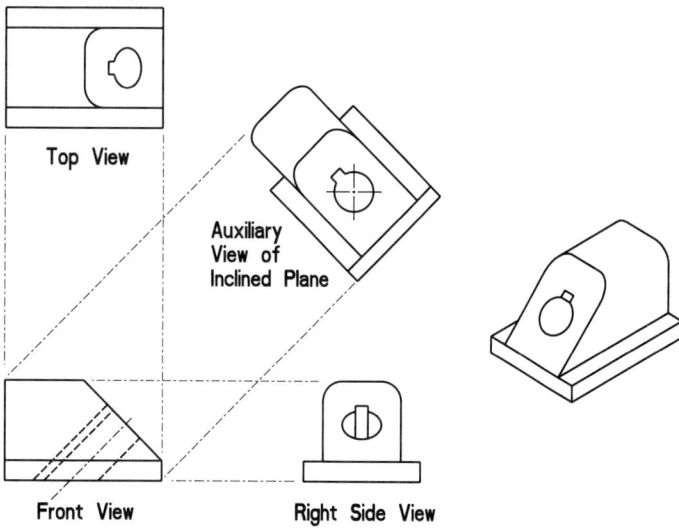

Figure 2.7.

2.4 WORKING DRAWINGS

Several types of *working drawings* are produced during the design process. Initially *freehand sketches* are used in the ideation phase of the design process. These are usually hand-drawn pictorial sketches of a concept that provide little detail, but enough visual information to convey the concept to other members of the design team. An example is the isometric sketch of a sheet metal piece that holds the blade of a pizza cutter, shown in Figure 2.8. The general shape of the object is clear, although details such as the thickness of the sheet metal and the radius of the bends in the sheet metal are not included. These conceptual sketches eventually evolve to final detailed drawings that define enough detail and information to support production.

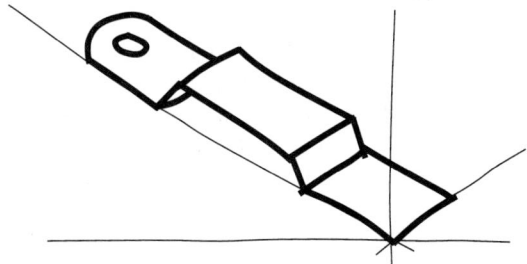

Figure 2.8.

Detail drawings document the detailed design of individual components using orthographic views. The detail drawing is the final representation of a design that is specific enough so that all of the information necessary for the manufacture of the part is provided. As a result, it is imperative that it includes the necessary views, dimensions, and specifications required for manufacturing the part. Figure 2.9 shows an example of a detail drawing of the part of the pizza cutter that was sketched in Figure 2.8. The detail drawing includes fully dimensioned orthographic views, notation of the material that the part is to be made from, information on the acceptable tolerances for the dimensions, and a title block that records important information about the drawing. Often an isometric view is included in the detail drawing to further clarify the shape of the part. Detail drawings provide sufficient detail so that the part can be manufactured based on the drawing alone.

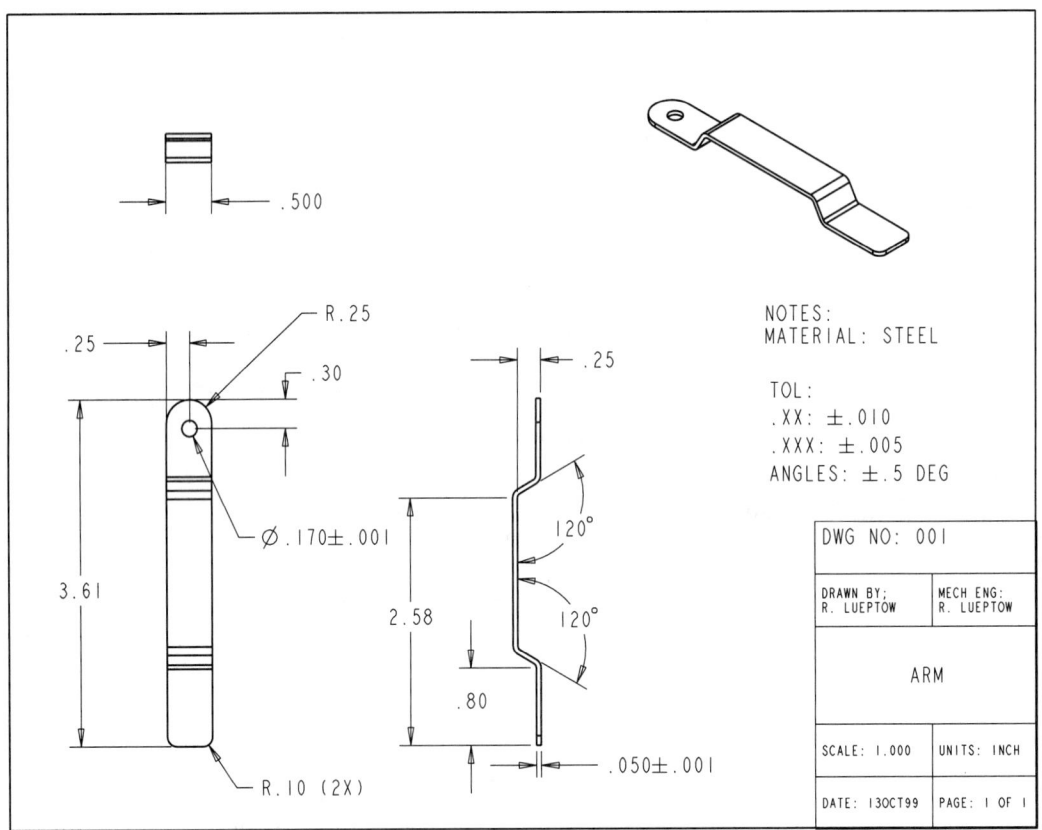

Figure 2.9.

Assembly drawings show how the components of a design fit together. Dimensions and other details are usually omitted in assembly drawings to enhance clarity. Several styles of assembly drawings are commonly used. Sometimes the assembly drawing is just an isometric view of the fully assembled device. But an exploded isometric view is often helpful to show the individual parts are assembled, as shown in Figure 2.10 for a pizza cutter. In some cases, a sectioned assembly, or cut-away view, shows how complicated devices are assembled. A cutting plane passes through the assembly and part of the device is removed to show the interior of the assembly. Numbers or letters can be assigned to individual parts of the assembly on the drawing and keyed to a parts list.

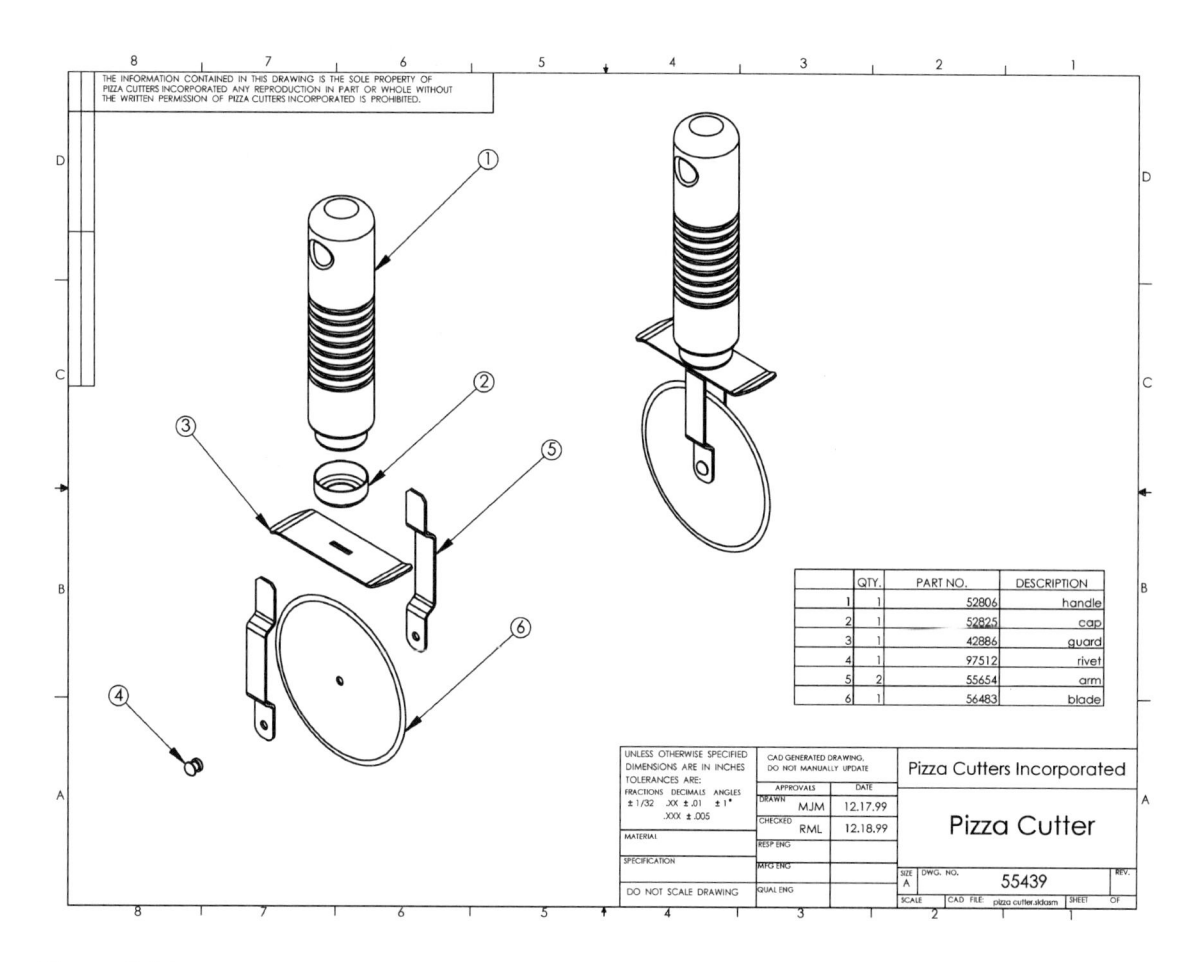

Figure 2.10.

Finally a *parts list*, or *bill of materials*, must be included with a set of working drawings. The parts list includes the part name, identification number, material, number required in the assembly, and other information (such as catalog number for standard parts such as threaded fasteners). An example is shown in Figure 2.11 for a pizza cutter. The parts list is used to ensure that all parts are ordered or manufactured and brought to the central assembly point.

	QTY.	PART NO.	DESCRIPTION
1	1	52806	handle
2	1	52825	cap
3	1	42886	guard
4	1	97512	rivet
5	2	55654	arm
6	1	56483	blade

Figure 2.11.

Taken together, the detail drawings of each individual part, the assembly drawing, and the bill of materials provide a complete set of working drawings for the manufacture of a part.

Chapter 2 Projections Used in Engineering Graphics

KEY TERMS

Assembly drawings
Bill of materials
Detail drawings
First-angle projection
Freehand sketches
Isometric projection
Multiview projection
Oblique projection
Orthographic projection
Perspective projection
3-D projections
Third-angle projection
Trimetric projection
Working drawings

Problems

1. Describe or sketch the front view that should be used in an orthographic projection of:

   a. a stapler.
   b. a television set.
   c. a cooking pot.
   d. a hammer.
   e. a pencil.
   f. a bicycle.
   g. an evergreen tree.
   h. a paper clip.
   i. a coffee mug.
   j. a padlock.

2. Identify the views shown in Figure 2.12 as isometric, trimetric, or perspective.

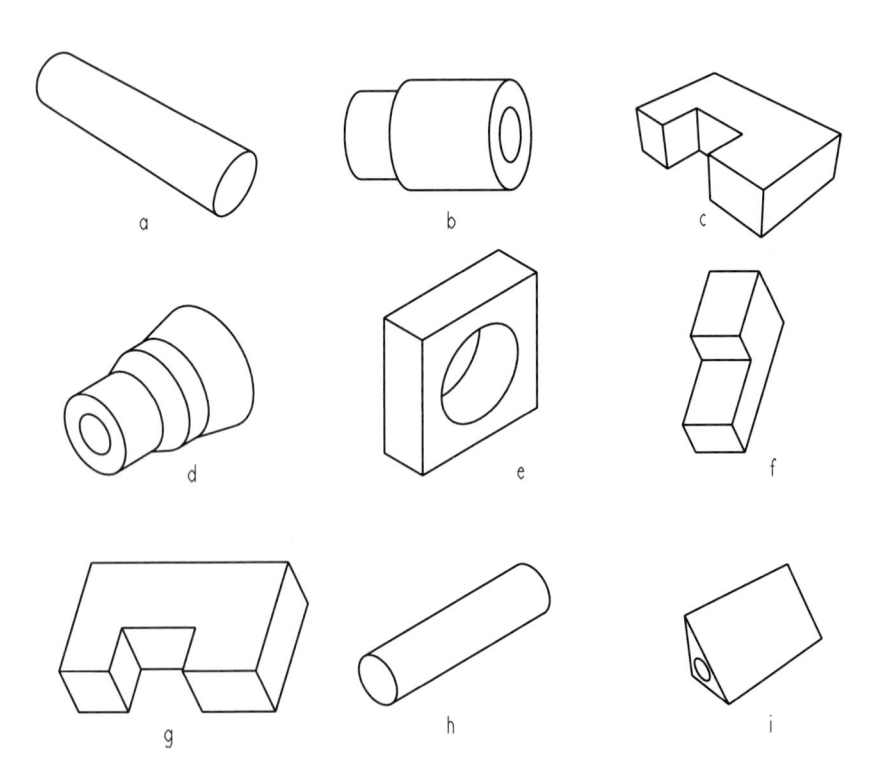

Figure 2.12.

3. For the drawings shown in Figure 2.13, determine whether the multiview projection is first-angle or third-angle.

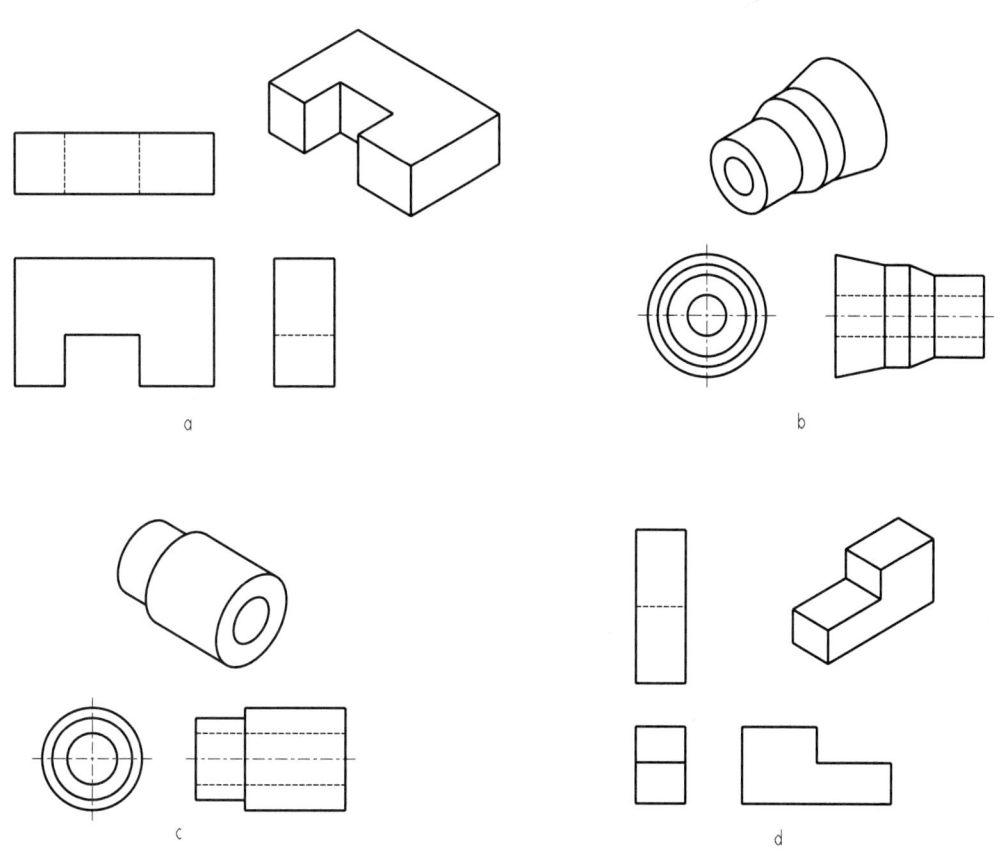

Figure 2.13.

4. Develop a bill of materials for:

   a. a pencil.
   b. a squirt gun.
   c. a click-type ball point pen.
   d. a videocassette (take an old one apart).
   e. an audio cassette (take an old one apart).
   f. a disposable camera (ask a local photo developer for a used one to take apart).
   g. eyeglasses.
   h. a household cleaner pump bottle.
   i. a claw-type staple remover.
   j. an adhesive tape dispenser.
   k. a bicycle caliper brake.
   l. a floppy disk (take an old one apart).
   m. a utility knife.
   n. a Vise-Grip wrench.

3

Freehand Sketching

OVERVIEW

In this chapter you will learn useful techniques for freehand sketching to create both two-dimensional orthographic sketches and three-dimensional pictorial sketches. You will learn how to quickly make rough sketches to convey a concept and how to make more refined sketches of objects that are more complex.

3.1 WHY FREEHAND SKETCHES?

An integral part of the creative design process is *ideation*, the generation of concepts or ideas to solve a design problem. Often freehand sketching can be used to explore and communicate mental concepts that come about in the mind's eye. The process of sketching can solidify and fill out rough concepts. Furthermore, sketching captures the ideas in a permanent form that can be used to communicate the concept to others. In this way, sketches often act as stepping stones to refine and detail the original concept or generate new ideas. Many great design ideas are first sketched on the back of an envelope or in a lab notebook, such as the freehand sketch of one of helicopter inventor Igor Sikorsky's designs (Figure 3.1.)

While computers are the workhorses for engineering graphics, initially generating ideas on a computer screen is very rare. A more common scenario is sketching an idea on paper and subsequently refining the concept on paper using more rough sketches. This often occurs simply because all

SECTIONS

- 3.1 Why Freehand Sketches?
- 3.2 Freehand Sketching Fundamentals
- 3.3 Basic Freehand Sketching
- 3.4 Advanced Freehand Sketching

OBJECTIVES

After reading this chapter, you should be able to:

- Explain why freehand sketching is important in design
- Freehand sketch lines and circles
- Sketch an oblique 3-D projection
- Sketch an isometric 3-D projection
- Sketch an orthographic multiview projection

that is needed for a freehand sketch is a pencil and a paper. Freehand sketching *quickly* translates the image of the concept in the mind's eye to paper. Engineers often communicate via rough freehand sketches to refine and improve the design. Sketches are much more useful than detailed CAD drawings early in the design process, because they are informal, quickly and easily changed, and less restrictive. It is only after clarifying the design concept by iterating through several freehand sketches that it is possible to draw the object using computer graphics. In fact, often an engineer will sit down to create a CAD drawing of an object using a freehand sketch as a guide.

This chapter focuses on the rudimentary elements of freehand technical sketching, because in many ways freehand sketching is the first step in CAD.

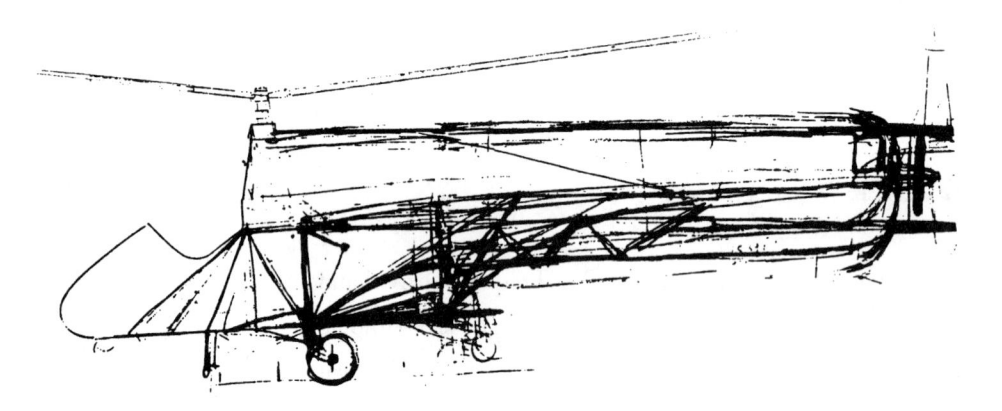

Figure 3.1. Helicopter inventor Igor Sikorsky's sketch of an early helicopter prototype demonstrates the visual impact of freehand sketching. (Used with permission of the Sikorsky Aircraft Corporation, Stratford, CT. ["Straight Up," by Curt Wohleber, *American Heritage of Invention and Technology,* Winter, 1993, pp. 26–39.])

3.2 FREEHAND SKETCHING FUNDAMENTALS

Freehand sketching requires few tools: just a pencil and paper. It may be tempting to use straight-edged triangles or rulers for drawing straight lines and a compass to draw circles. But these instruments often slow down the process and distract from the purpose of sketching, which is to create a quick, rough graphical representation of the image in the mind's eye. Generally sketching has three steps, although the steps are usually subconscious. First, the sketch is planned by visualizing it in the mind including the size of the sketch on the paper, the orientation of the object, and the amount of detail to be included in the sketch. Second, the sketch is outlined using very light lines to establish the orientation, proportion, and major features of the sketch. Finally, sharpening and darkening object lines and adding details develops the sketch.

All sketches are made up of a series of arcs and lines, so the ability to draw circles and straight lines is necessary. A straight line is sketched in the following way. First, sketch the endpoints of the line as dots or small crosses. Then place your pencil on the starting endpoint. Keeping your eyes on the terminal point, use a smooth continuous stroke to draw the line between the points as shown in Figure 3.2. Nearly horizontal or vertical lines are frequently easier to draw than inclined lines, so it may be helpful to shift the paper to draw the line horizontally or vertically. For long lines, it may be helpful to mark two or three points along the line and use the procedure between consecutive points or to make two or three shorter passes lightly with the pencil before a final darker line.

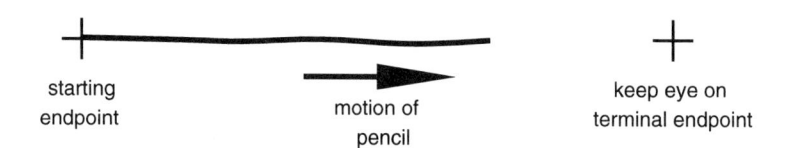

starting endpoint   motion of pencil   keep eye on terminal endpoint

Figure 3.2.

A circle can be sketched using the following steps, illustrated in Figure 3.3a. First, draw light horizontal and vertical lines crossing at the center of the circle. Second, lightly mark the radius of the circle on each line. Finally, connect the radius marks with a curved line to form a circle. Another technique is to lightly draw a square box the same size as the circle diameter as shown in Figure 3.3b. Then lightly draw diagonals of the box and centerlines between midpoints of the sides of the box. The diagonals and centerlines should intersect at the center of the circle. Mark the radius on these lines, and sketch the circle within the box. It is sometimes helpful to mark the radius on the edge of a scrap paper and mark the radius at as many points as desired in addition to the marks on the centerlines and diagonals. Arcs are sketched in much the same way as circles, except that only a portion of the circle is sketched. It is generally easier to sketch an arc with your hand and pencil on the concave side of the arc.

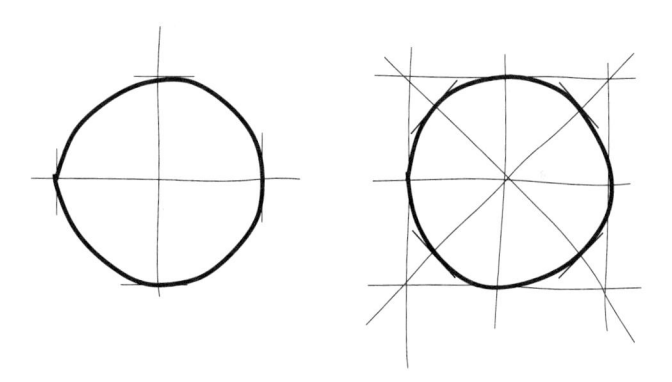

Figure 3.3.

3.3 BASIC FREEHAND SKETCHING

Many times, particularly during the conceptual stage of design, it is necessary to immediately communicate a graphical image to others. It has been said that some of the best design engineers are the ones who can sketch an idea clearly in a minute or so. The goal of the sketch in this case is not to show the details of the part, but to provide another person with a clear concept of the idea. For example, a design engineer may need to show a sketch to a manufacturing engineer to get input on the manufacturability of a part. If the concept is at an early phase, CAD drawings would not have been created yet. So the design engineer needs to use a freehand sketch of the part.

The sketch of Sikorsky's helicopter in Figure 3.1 exemplifies the power of freehand sketching. A brief glance at this sketch provides immediate insight to the concept that is being shown. One does not need to study the sketch to know what is being sketched, even if the viewer has never seen the concept before. These quick ideation sketches are not difficult to draw and require no artistic talent, just some practice.

Two types of pictorial sketches are used frequently in freehand sketching: oblique and isometric. The *oblique projection* places the principal face of the object parallel to the plane of the paper. The *isometric projection* tilts the part so that no surface of the part is in the plane of the paper. The advantage of the oblique projection is that details in the front face of the object retain their true shape. This often makes oblique freehand sketching easier than isometric sketching, where no plane is parallel to the paper. The disadvantage of the oblique projection is that it does not appear as "photorealistic" as an isometric projection. In other words, an isometric projection is similar to what a photograph of the object would look like.

3.3.1 Oblique Sketching

Often freehand sketching begins with light thin lines called *construction lines* that define enclosing boxes for the shape that is being sketched. Construction lines are used in several ways. First, the construction lines become the path for the final straight lines of the sketch. Second, the intersection of construction lines specify the length of the final lines. Third, points marked by the intersection of construction lines guide the sketching of circles and arcs. And finally, construction lines guide the proportions of the sketch. This last item is of crucial importance if the sketch is to clearly represent the object. For example, if an object is twice as wide as it is high, the proportions in the sketch must reflect this. Proper proportions of the boxes defined by the construction lines will result in proper proportions of the sketch.

An oblique freehand sketch is easy, since it begins with a two-dimensional representation of the face of the object. Figure 3.4 shows the steps in quickly sketching a part with a circular hole.

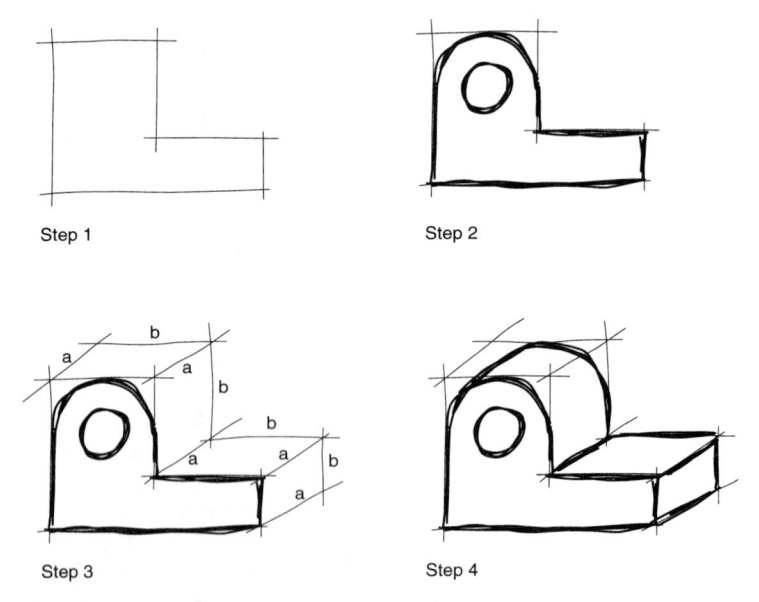

Figure 3.4.

Step 1: Horizontal and vertical construction lines are lightly drawn to outline the basic shape of the main face of the part. This is known as *blocking-in* the sketch. If you are using a pencil or felt-tip marker, press lightly when drawing the construction lines to produce a thin or light line. If you are using a ball-point pen, draw a single, light line.

Step 2: Sketch in the face of the part using the construction lines as a guide. How you sketch the outline of the part depends on the type of pen or pencil that you are using. The idea is to thicken the lines of the part compared to the construction lines. If you are using a pencil or a felt-tip marker, pressing hard for the outline of the part will result in heavy or dark lines. If you are using a ball-point pen, the line width does not depend much on how hard you press. In this case, the outline of the part is sketched with a back and forth motion of the pen to thicken the lines of the part compared to the construction lines as shown in Figure 3.4. The straight lines are usually sketched first, followed by the arcs. The circle for the hole in the part is added last to complete the face of the part.

Step 3: Sketch *receding construction lines* (lines into the plane of the paper labeled *a*) at a convenient angle. All of the receding lines must be parallel to each other and are usually at an angle of 30° to 45°. The receding lines end at the appropriate depth for the object. Then vertical and horizontal lines at the back plane of the part are added (lines labeled *b*). This blocks in the three-dimensional box enclosing the object.

Step 4: Sketch in and darken the lines outlining the part. Again it is usually easiest to sketch in the straight lines first, then the arcs, and finally any details. Because the construction lines are light compared to the outline of the part, they are not erased.

The final sketch, while rough and lacking detail, clearly shows the design intent for the part.

3.3.2 Isometric Sketching

Isometric freehand sketches are somewhat more difficult to master than oblique sketches because no face is in the plane of the paper in an isometric view. The steps to construct a simple freehand isometric sketch are shown in Figure 3.5.

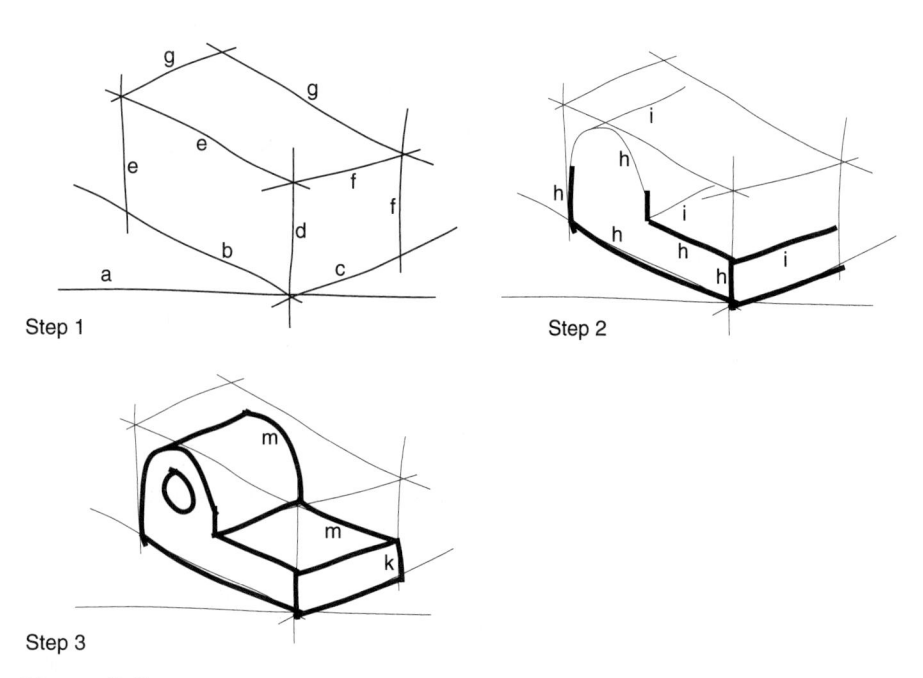

Figure 3.5.

Step 1: Sketch a light horizontal line (*a*). From this line draw two intersecting lines at an angle of approximately 30° to the horizontal (*b* and *c*). Then draw a vertical line (*d*) through the intersection of the previous three lines. The three lines labeled *b*, *c*, and *d* form the isometric axes of the sketch. Next sketch the box to block in the front face of the part (*e*). These lines should be parallel to axes *b* and *d*. Similarly, sketch the lines to block in the right face (*f*) making sure that the lines are parallel to axes *c* and *d*. Finish this step by sketching lines parallel to the axes to complete the box that encloses the part (*g*).

Step 2: The outline for the front face is added by sketching in lines and curves (*h*). Then outline the front face using heavy lines. In this case, a single heavy line such as might be produced from pressing hard on a pencil or felt-tip marker is used. Next, lines are sketched to indicate the depth of the features of the front view (*i*). These lines should be parallel to axis *c*. They can be darkened after they are drawn lightly.

Step 3: Finally, a line is added to complete the back corner of the part (*k*). Lines and arcs are added to complete the back face of the part (*m*). Then the hole detail is added. Circular holes appear as ellipses in isometric views, as discussed in the next section.

The choice of whether to use an oblique projection or an isometric projection is often arbitrary. Because the oblique projection is easier to sketch, it is sometimes preferred. On the other hand, an isometric projection provides a more photorealistic image of the object.

3.4 ADVANCED FREEHAND SKETCHING

The sketching methods described in the previous section were focused on sketches in which the face of the object is in a single plane. Freehand sketching is somewhat more difficult when the face of the object is not in a single plane. The difficulty here is accurately depicting the depth of the object. Oblique and isometric projections are still useful, though somewhat more complicated than those in the previous section. In addition, orthographic projections are also valuable.

3.4.1 Freehand Oblique Sketching

An example of the steps leading to an oblique freehand sketch of a complicated object are shown in Figure 3.6. Because the face of the base of the object and the face of the upper portion of the object are in different planes, it is necessary to begin with a box that encloses the entire object before sketching either face. Some of the construction lines are removed after they are used in this example. This was done here to make the sketch more clear. However, this is not necessary in practice, if the construction lines are drawn as light lines.

Step 1: To begin, construction lines to form a box that encloses the object are drawn to block-in the sketch. Notice that the front and back faces of the box are rectangular with horizontal and vertical sides. The receding construction lines are parallel and at an angle of 30° to 45° to horizontal. The easiest way to draw this box is to first draw the front rectangle (*a*). Then draw an identical second rectangle above and to the right of the first rectangle (*b*). Finally connect the corners with receding construction lines (*c*).

Step 2: Now the front face of the base of the object can be sketched in the front rectangle. The lines are appropriately darkened.

Section 3.4 Advanced Freehand Sketching   29

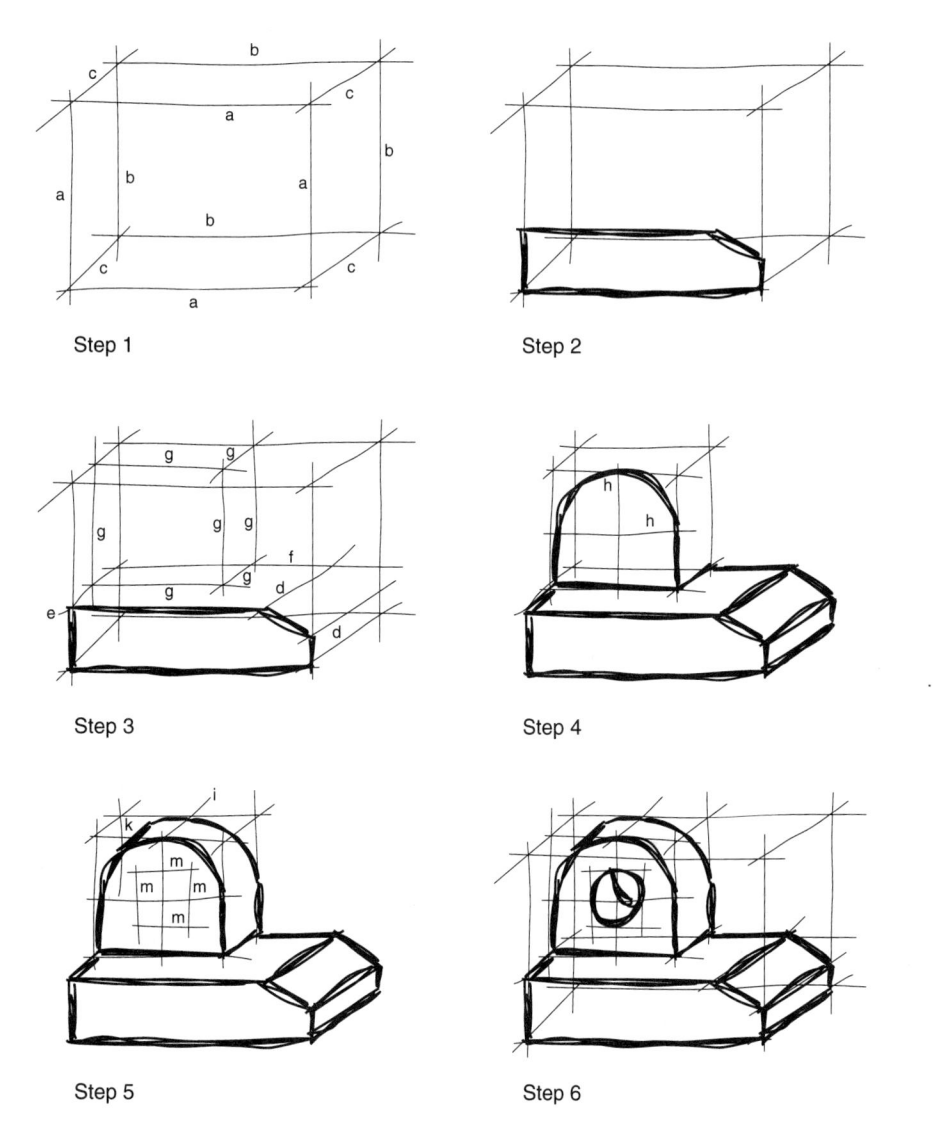

Step 1

Step 2

Step 3

Step 4

Step 5

Step 6

Figure 3.6.

Step 3: Certain features of the front face of the base extend backward along or parallel to the receding construction lines. For example, the lines (*d*) forming the chamfer (angled cut on the right side of the base) can be sketched parallel to receding lines. Likewise the receding line for the upper left corner of the base can be sketched (*e*). Then the base can be finished with a horizontal line on the back face (*f*). Now it is possible to block in the upper rounded portion of the object to create a box (*g*) that encloses the upper protrusion within the larger box that encloses the entire object.

Step 4: The front face of the upper portion of the object can be sketched in this box. Then receding lines corresponding to the chamfer and the left edge of the base can be darkened. In addition, the lines forming the back face can be sketched. Note that the line forming the back edge of the chamfer is parallel to the line forming the front edge of the chamfer. Construction lines (*h*) on the front face of the upper portion are drawn to center of the circle for the hole.

Step 5: A receding construction line (*i*) extending from the peak of the front face to the plane of the back face is sketched to aid in aligning the curved outline of the back of the upper portion. The back face is identical to the front face except that it is shifted upward and to the right. This results in the left side of the back face being hidden. A darkened receding line (*k*) finishes the left side of the upper portion of the object. Finally, four construction lines (*m*) are sketched to block in the circle for the hole.

Step 6: Now the hole can be sketched in and darkened. The back edge of the hole is also added to complete the sketch. The construction lines may be erased, but usually the construction lines are retained if they are made properly as light lines.

Oblique sketching is often aided by the use of graph paper with a light, square grid. The process is identical to that shown in Figure 3.6, but it is easier to keep the proportions correct by counting the number of boxes in the grid to correspond to the approximate dimensions of the part. Graph paper further improves the sketch by helping keep lines straight as well as more accurately horizontal or vertical.

3.4.2 Isometric Sketching

Isometric freehand sketches of more complex objects start with an isometric box to block in the sketch. Then faces are sketched and additional features are blocked in. Finally details are added. The steps to construct an isometric sketch are shown in Figure 3.7. Some of the construction lines are removed after they are used, to make the sketch more clear in this figure. Normally, removing construction lines is not necessary.

Step 1: To begin, sketch a light horizontal line (*a*). From this line draw two intersecting lines at an angle of approximately 30° to the horizontal (*b* and *c*) and a vertical line (*d*) through the intersection of the previous three lines to form the isometric axes of the sketch. Finish blocking in by sketching lines (*e*) to complete the box so that it will completely enclose the object. Unlike the oblique sketch, it is often better not to sketch hidden construction lines when blocking in.

Step 2: Block in the front face of the part (*f*) so that the construction line is parallel to the isometric axis. Similarly, sketch the line to block in the right face (*g*).

Step 3: Sketch the left face and the right face and darken the lines. This completes the faces that are in the front planes of the box. Now sketch in three lines (*h*) parallel to the isometric axis (*c*). The left line (*h*) is the top edge of the base. The middle line (*h*) finishes the chamfer. The right line (*h*) is used to aid in sketching a construction line for the back edge of the base (*i*), which is sketched next.

Step 4: Now the face of the chamfer can be darkened and the angled line at the back edge of the chamfer can be added. This completes the angled face of the chamfer. Next the protrusion above the base can be blocked in with seven lines (*k*).

Step 5: The front face of the upper protrusion is sketched first using light lines. Construction lines (*m*) are added to help identify the location of the endpoints of the arc of the front and back faces of the protrusion. The rounded rear face (*n*) is sketched lightly to be identical to the front face, except that part of it is not visible. The line at the top left edge of protrusion (*o*) is added. Then all lines forming the upper portion of the object are darkened In addition, the line forming the top edge of the base on the back side is darkened.

Step 6: The details related to the hole are added next. Circles in isometric projections are difficult to draw because they appear as ellipses with their major axes at an angle to horizontal. The center of the hole is where two lines (m) intersect on the front face of the upper portion of the object. The lines (p) forming the parallelogram to enclose the ellipse for the hole are added. Each side of the parallelo-

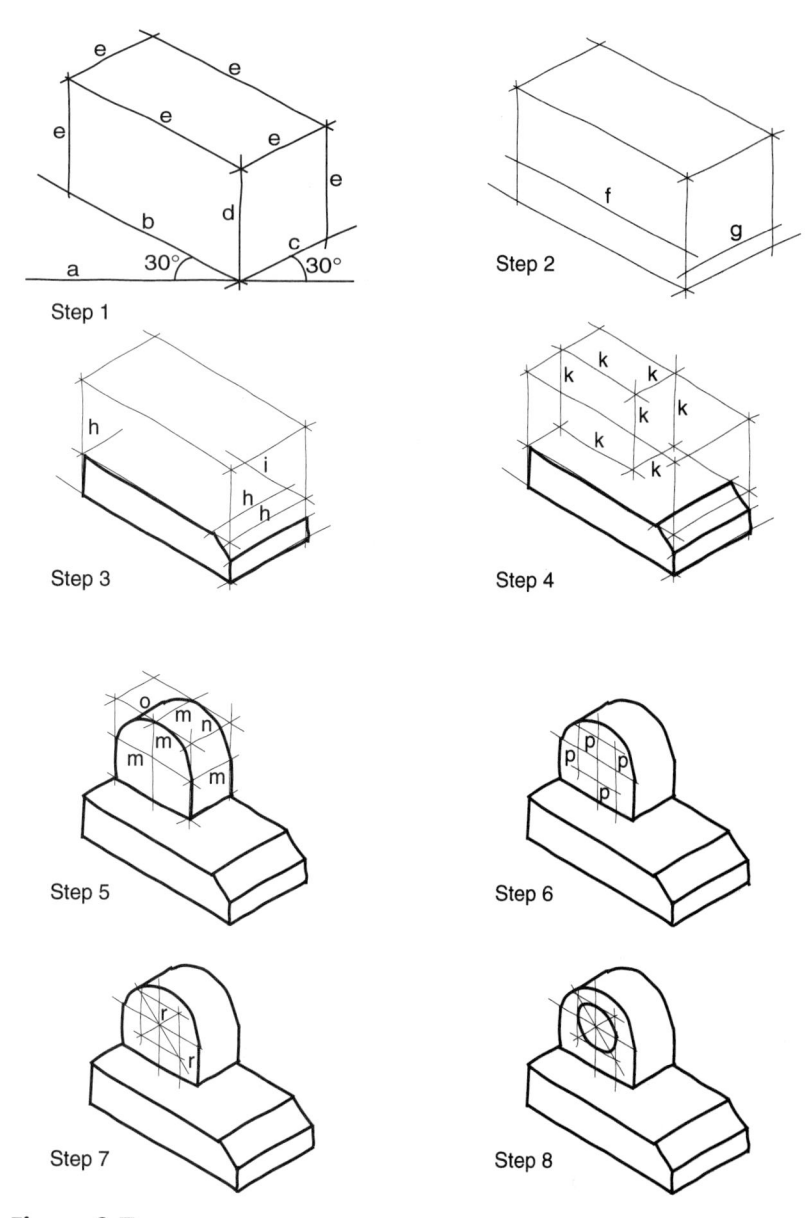

Figure 3.7.

gram should be parallel to one of the isometric axes. The sides should be equal in length to one another.

Step 7: To help in sketching the ellipse, construction lines forming the diagonals of the parallelogram (*r*) are added.

Step 8: Now the ellipse that represents the circular hole can be sketched. A few simple points help in sketching ellipses more easily. The major axis and minor axis of the ellipse are perpendicular to one another. The major and minor axes also coincide with the diagonals of the parallelogram enclosing the ellipse (*r*). The ellipse touches the parallelogram at the midpoints of the sides of the parallelogram. Start drawing the hole by sketching short elliptical arcs between the midpoints of the parallelogram on either side of the minor axis. Finish the hole by sketching sharply curved elliptical arcs between the midpoints of the parallelogram on either side of the major axis of the hole. Finally, darken and make heavy the lines outlining the hole and any remaining edges of the part.

Isometric sketching is made substantially easier by the use of isometric grid paper. This paper has a grid of lines at horizontal and 30° to horizontal (corresponding to lines b, c, and d in Figure 3.7). The procedure for using isometric grid paper is the same as that described above, but using the isometric grid paper keeps proportions of the part consistent. One simply counts grid boxes to approximate the dimensions of the object. The grid paper also aids in sketching straight lines parallel to the isometric axes.

3.4.3 Orthographic Sketching

In some cases it is necessary to sketch orthographic projection views rather than oblique or isometric pictorial views. Because orthographic views are two-dimensional representations, they are not as difficult to sketch as pictorial views. But there are several techniques that make freehand sketching of orthographic views easier and more efficient. The process for sketching three orthographic views of the object in the previous two figures is shown in Figure 3.8.

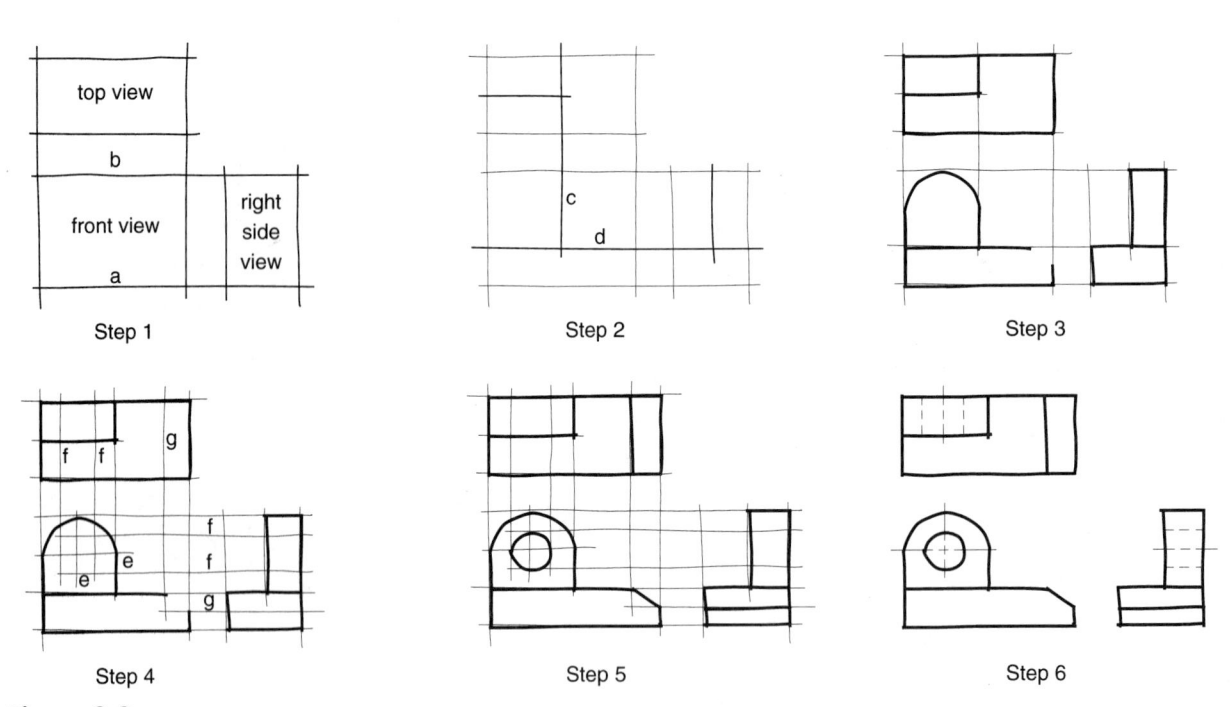

Figure 3.8.

Step 1: Begin by blocking in the front, top, and side views of the object using the overall width, height, and depth. The construction lines extend between views to properly align the views and maintain the same dimension in different views. For instance, line (*a*) represents the bottom edge and line (*b*) represents the top edge in both the front view and the right-side view. The distance between lines (*a*) and (*b*) is the height dimension in both views. The space between the views should be large enough so that the drawing does not look crowded and should be the same between all views.

Step 2: The upper protrusion is blocked in. Note that line (*c*) extends across the top and front views, to assure that the width of the protrusion is consistent in both views. Likewise, line (*d*) extends across the front and right-side views.

Step 3: The outline of the object is darkened to clearly show the shape of the object in all three views. Care must be taken in darkening lines. For instance, the right corner of the front view should not be darkened, because the detail of the chamfer has not yet been added.

Step 4: Construction lines for the holes and other details are added next. The center of the hole is positioned with construction lines (*e*). Then construction lines that block in the hole (*f*) are drawn. These construction lines extend between views to project the hole to the top view and to the right-side view. Construction lines extending between views (*g*) are also added for the chamfer.

Step 5: Now the hole and chamfer are sketched and darkened to show the completed object.

Step 6: Finally, centerlines (long-dash, short-dash) that indicate the center of the hole are added. Hidden lines (dashed lines) that indicate lines hidden behind a surface are also added. Construction lines may be erased as was done in this figure, but this is not usually necessary.

The quality of the sketch can often be improved by using square grid graph paper to keep proportions and act as a guide for horizontal and vertical lines. Some engineers prefer to use a straight-edge to produce a nicer sketch, but this is usually not necessary with practice and sufficient care in sketching.

PROFESSIONAL SUCCESS

Will Freehand Sketching Ever Become Obsolete?

CAD has almost totally eliminated pencil and paper drawings. But what about pencil and paper freehand sketching? Although many computers and palm-tops offer sketching programs, it is unlikely that freehand sketching will disappear soon. Just as it is easier to do a calculation in your head or on a piece of scratch paper rather than finding a calculator and punching in the numbers, it is easier to sketch an image on a piece of paper (or a napkin!) than to find a computer, log in, and start the appropriate "paint" program. Pencil and paper freehand sketches are quick, efficient, easily modified, and easily conveyed to others. And all that is needed is a pencil and a scrap of paper.

Even if the pencil and paper are totally replaced by palm-top computers some day, freehand sketching skills will still be useful. Instead of using a pencil on a piece of paper, we will use a stylus on a touch screen. The only difference is the medium. The freehand sketching techniques themselves are unlikely to change much. Perhaps one could imagine sketching software that assists in generating oblique or isometric sketches as they are drawn at some time in the future.

KEY TERMS

Blocking-in
Construction lines
Ideation
Receding construction lines

Problems

For all of the problems, the items shown in Figure 3.9, 3.10, and 3.11 are 2 inches wide, 1.5 inches high, and 1 inch deep. The holes in Figure 3.9b, 3.9c, 3.9d, 3.10c, and 3.10d are through holes. The hole in Figure 3.10b is through the front face only.

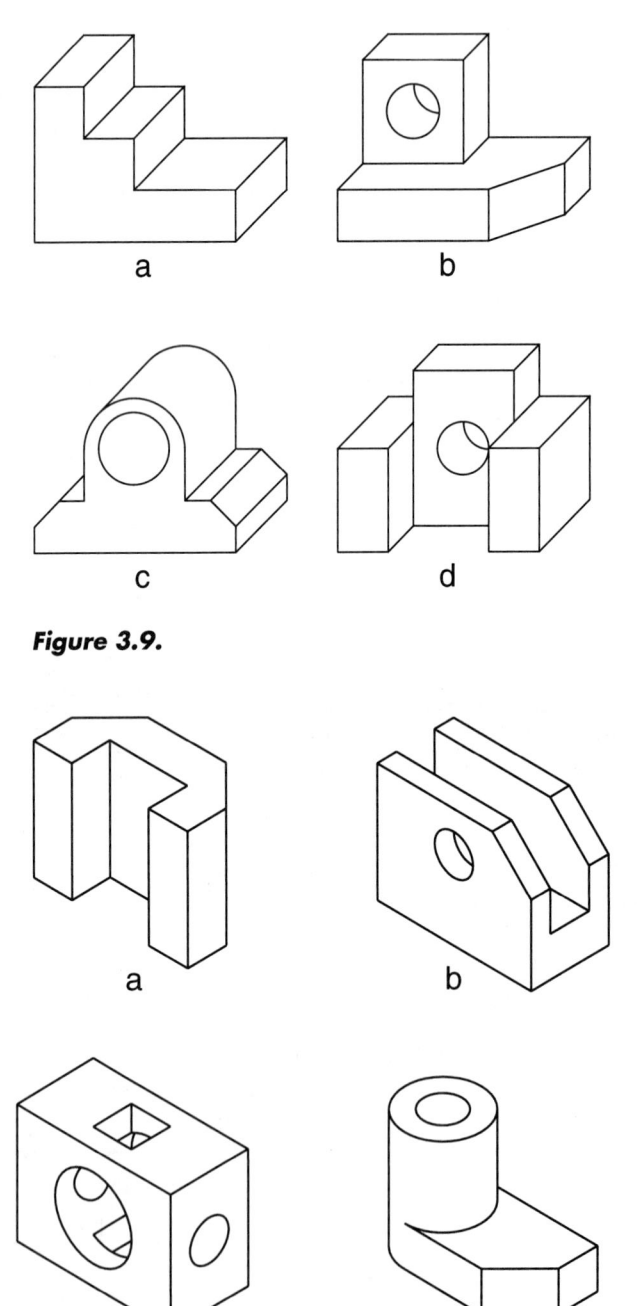

a

b

c

d

Figure 3.9.

a

b

c

d

Figure 3.10.

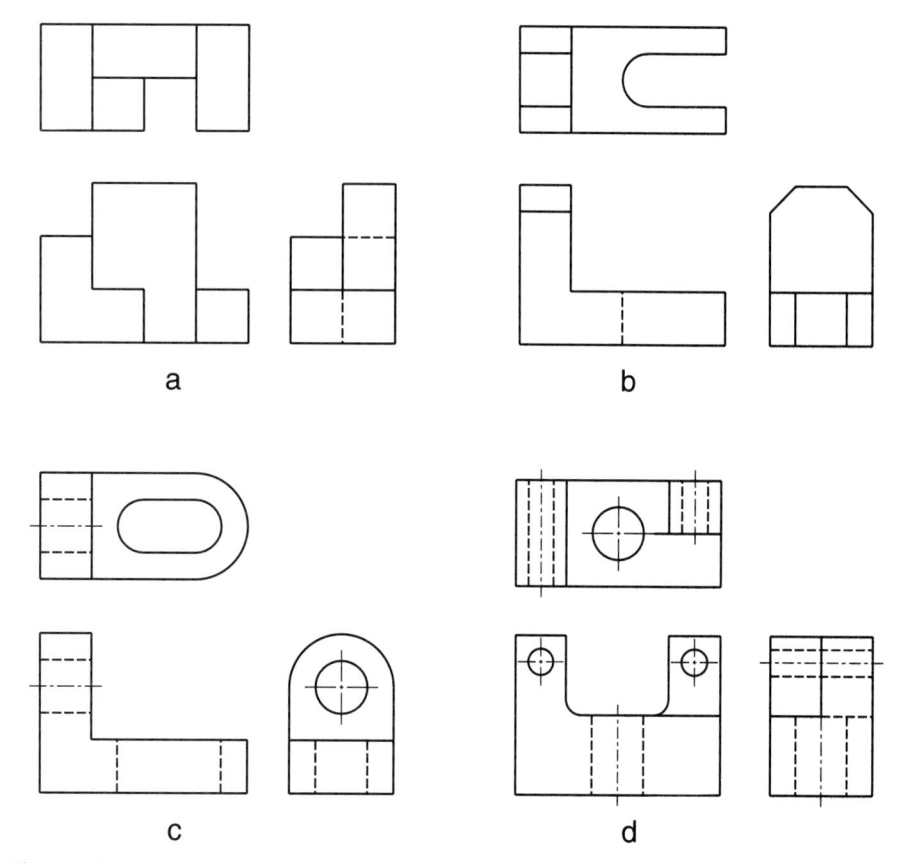

Figure 3.11.

1. Create freehand oblique sketches of the objects in Figure 3.9. (The objects are shown as oblique projections, so you must simply recreate the drawing by freehand sketching.)
2. Create freehand oblique sketches of the objects in Figure 3.10. (The objects are shown as isometric projections.)
3. Create freehand oblique sketches of the objects in Figure 3.11. (The objects are shown as orthographic projections.)
4. Create freehand isometric sketches of the objects in Figure 3.10. (The objects are shown as isometric projections, so you must simply recreate the drawing by freehand sketching.)
5. Create freehand isometric sketches of the objects in Figure 3.9. (The objects are shown as oblique projections.)
6. Create freehand isometric sketches of the objects in Figure 3.11. (The objects are shown as orthographic projections.)
7. Create freehand orthographic sketches of the objects in Figure 3.11. (The objects are shown as orthographic projections, so you must simply recreate the drawing by freehand sketching.)
8. Create freehand orthographic sketches of the objects in Figure 3.9. (The objects are shown as oblique projections.)
9. Create freehand orthographic sketches of the objects in Figure 3.10. (The objects are shown as isometric projections.)

4

Computer Aided Design and Drafting

Overview

Several different models for representing a part's geometry have been used in CAD, including 2-D models and 3-D wireframe and surface models. However, solids modeling is the current state-of-the-art in CAD. Solids modeling has the inherent advantage of more accurately representing the "design intent" of the part that modeled.

4.1 CAD MODELS

A CAD *model* is a computer representation of an object or part. It can be thought of as a "virtual" part in that it exists only as a computer image. The model is an engineering document of record. It contains all of the design information including geometry, dimensions, tolerances, materials, and manufacturing information. A CAD model replaces the paper blueprints and engineering drawings of just a few decades ago.

The simplest model used in CAD is a *2-D model*. This model is essentially the computer graphics equivalent to an orthographic projection created using a pencil and paper as shown in Figure 4.1. A 2-D model represents a three-dimensional object with several views, each showing the view of one side of the object projected onto a plane. On the computer, a 2-D model appears as lines and curves on a flat surface, and it is up to the engineer or designer to interpret several views to create a mind's eye image of the three-dimensional object. The CAD software electronically stores

Sections

- 4.1   CAD Models
- 4.2   CAD and Solids Modeling
- 4.3   The Nature of Solids Modeling

Objectives

After reading this chapter, you should be able to:

- Differentiate 2-D, 2 1/2-D, and 3-D models
- Differentiate wireframe, surface, and solids models
- Explain how a cross-section is extruded or revolved to create a solids model
- Explain the analogy between creating a solids model of a part and machining the part
- Explain feature-based modeling
- Explain constraint-based modeling
- Explain associative modeling

the 2-D model as several views but does not inherently "know" that the views can be connected to one another to form a three-dimensional object.

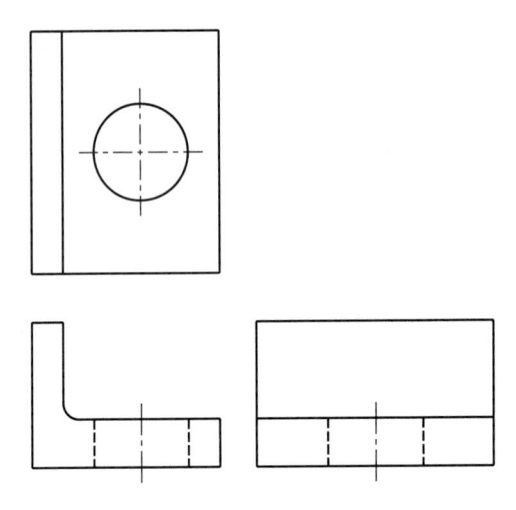

Figure 4.1.

A *2 1/2-D model* has a third dimension that is recognized by the CAD software, but the third dimension is simply an extrusion of a two-dimensional shape. In this case, the CAD software electronically stores the cross-section shape of the object and the depth that it is *extruded*, or stretched, in the third dimension. Thus, the only objects that can be represented using a 2 1/2-D model model are those that have a constant cross-section. For instance, the simple shape shown in Figure 4.2 is the 2 1/2-D model of the object shown using a 2-D model in Figure 4.1 without the hole. The object, including the hole, cannot be represented as a 2 1/2-D model because the hole results in a change in the cross-section. A 2 1/2-D model could be used to represent a shape such as an I-beam or C-channel.

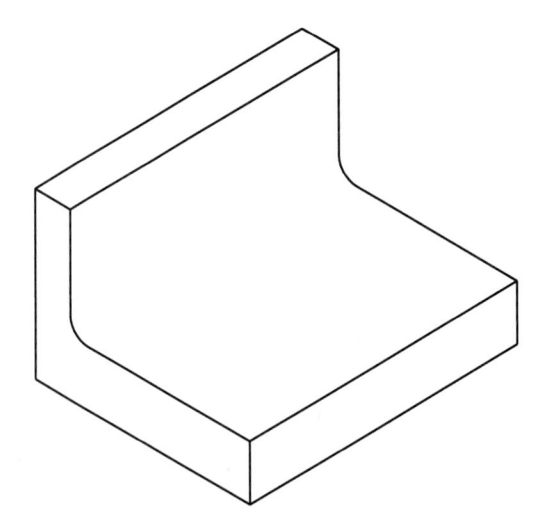

Figure 4.2.

A *3-D model* is the most general model used in CAD software. In this case, the CAD software electronically stores the entire three-dimensional shape of the object. Most simply, one can think of the software storing the three-dimensional coordinates of various points on the object and then defining how these points are connected. Of course, a two-dimensional drawing representing the orthographic views of the object can be automatically created from the 3-D model. The current generation of CAD software is based on 3-D models.

The simplest 3-D model is a *wireframe model*. Figure 4.3a shows a wireframe representation of the object that we have been considering. In a wireframe model, only edges of the object are represented. Thus, the CAD software needs to store the locations of the vertices (intersections of lines) and which vertices are to be connected to each other by lines. A circle or curved section can be represented by a series of closely spaced vertices with linear edges connecting them or as an arc defined by a center, radius, and end points. Before more powerful computers were available, the low storage requirements of a wireframe model made it quite popular. The problem inherent in a wireframe representation is quite clear in Figure 4.3a—it is difficult to interpret the drawing because all of the edges are visible. In fact, there are many situations where a wireframe representation cannot be unambiguously interpreted because it is not clear which lines would remain hidden if the surfaces of the part were filled in. This problem can be solved using various methods to remove "hidden lines" from the wireframe. This is tricky because the CAD software has to first determine how the edges of the wireframe form a surface and then determine if that surface hides any lines. Figure 4.3b shows the object with the hidden lines shown as dashed lines to indicate that they are behind a surface. 3-D models other than wireframes are able to deal with hidden lines more effectively and offer other advantages.

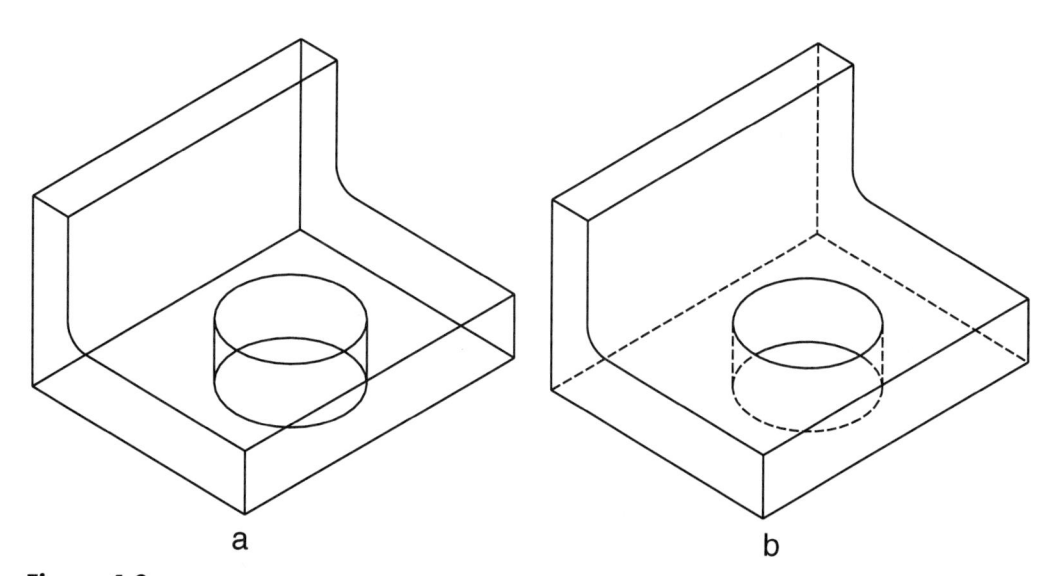

Figure 4.3.

A 3-D *surface model* defines the object in terms of surfaces such as plates (flat) and shells (curved) in addition to edges. This makes it easy to determine whether a line or surface is hidden, because the surfaces are defined. Using a surface model also permits the construction of smoothly curved surfaces, such as a circular hole, rather than using many line segments to form a curved line as with some wireframe models. It is also possible to "fair" or smooth one surface into another surface to provide a sculpted shape.

Figure 4.4a shows the simple object with hidden lines removed altogether. Although a surface model can be displayed to look like a wireframe, either with or without hidden lines, it can also be displayed by assigning different degrees of shading to the surfaces. A virtual light source is assumed to be near the object to provide a three-dimensional lighting effect as shown in Figure 4.4b. This is known as *shading* or *rendering*. Although a variety of shading techniques are available, all depend on determining how much light from a virtual light source strikes each portion of the virtual object. The use of shading provides a very realistic image of the object, which permits a much more vivid communication of the nature of the object. However, the problem that remains with a surface model is that the interior of the object remains undefined. The surface model just represents the shell of the object.

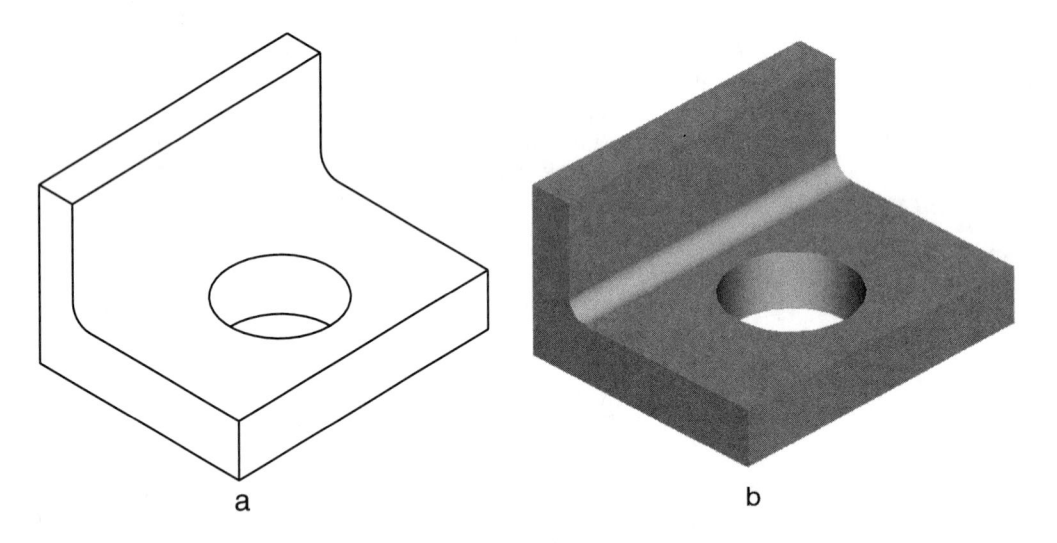

a   b

Figure 4.4.

4.2 CAD AND SOLIDS MODELING

Solids modeling, the current state-of-the-art in CAD, is the most sophisticated method of representing an object. Unlike wireframe or surface models, a solids model represents an object in the virtual environment just as it exists in reality, having volume as well as surfaces and edges. In this way, the interior of the object is represented in the model as well as the outer surfaces.

The first attempt at solids modeling was a technique known as *constructive solid geometry* or *primitive modeling*, which is based on the combination of geometric primitives such as right rectangular prisms (blocks), right triangular prisms (wedges), spheres, cones, and cylinders. Each primitive could be scaled to the desired size, translated to the desired position, and rotated to the desired orientation. Then one primitive could be added to or subtracted from other primitives to make up a complex object. It is easy to see how two blocks plus a negative cylinder (hole), as shown in Figure 4.5, could represent an object similar to the one shown in Figure 4.4. (The rounded surface, or fillet, at the corner between the horizontal and vertical surfaces has been omitted.) The problem with constructive solid geometry was that the mental process of creating a solid model based on geometric primitives was much more abstract than the mental processes required for designing real world objects.

Figure 4.5.

Constraint-based solids modeling overcomes the weakness of constructive solid geometry modeling by making the modeling process more intuitive. Instead of piecing together geometric primitives, the constraint-based modeling process begins with the creation of a 2-D sketch of the profile for the cross-section of the part. Here, "sketch" is the operative word. The sketch of the cross-section begins much like the freehand sketch of the face of an object in an oblique view. The only difference is that CAD software draws straight lines and perfect arcs. The initial sketch need not be particularly accurate; it need only reflect the basic geometry of the part's cross-sectional shape. Details of the cross-section are added later. The next step is to constrain the 2-D sketch by adding enough dimensions and parameters to completely define the shape and size of the 2-D profile. The name *constraint-based modeling* arises because the shape of the initial 2-D sketch is "constrained" by adding dimensions to the sketch. Finally, a three-dimensional object is created by *revolving* or *extruding* the 2-D sketched profile. Figure 4.6 shows the result of revolving a simple L-shaped cross-section by 270° about an axis and extruding the same L-shaped cross-section along an axis. In either case, these solid bodies form the basic geometric solid shapes of the part. Other features can be subsequently added to modify the basic solid shape.

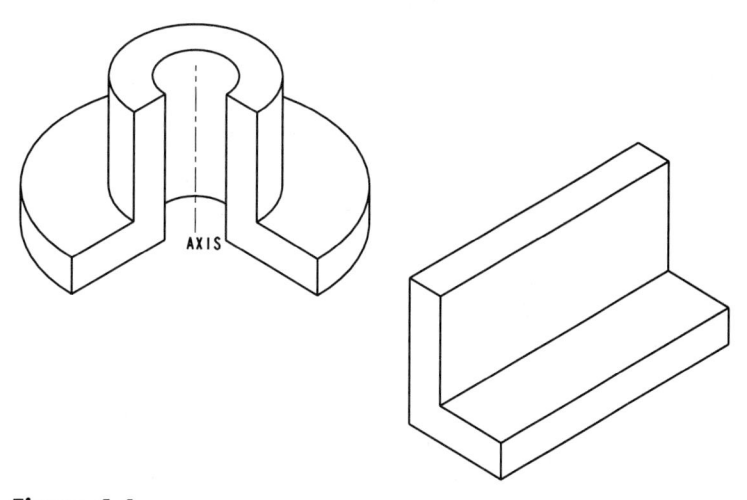

Figure 4.6.

Once the solids model is generated, all of the surfaces are automatically defined, so it is possible to shade it in the same way as a surface model is shaded. It is also easy to generate 2-D orthographic views of the object. This is a major advancement from other modeling schemes. With solids modeling, the two-dimensional drawings are produced as views of the three-dimensional virtual model of the object. In traditional CAD, the three-dimensional model is derived from the two-dimensional drawings. One might look at solids modeling as the sculpting of a virtual solid volume of material. Because the volume of the object is properly represented in a solids model, it is possible to slice through the object and show a view of the object that displays the interior detail. Once several solid objects have been created, they can be assembled in a virtual environment to make sure that they fit together and to visualize the assembled product.

PROFESSIONAL SUCCESS

How Solids Models are Used

Solids models are useful for purposes other than visualization. The solids model contains a complete mathematical representation of the object, inside and out. This mathematical representation is converted easily into specialized computer code that can be used for stress analysis, heat transfer analysis, fluid flow analysis, and computer-aided manufacturing.

Finite Element Analysis (FEA) is a method to subdivide the object under study into many small, simply shaped elements. The simple geometry of the finite elements allows the relatively simple application of the appropriate equations for the stress and deformation of the elements or for the heat transfer between elements. By properly relating nearby elements to one another, all of these equations for each element can be solved simultaneously. The final result is the stress, deformation (strain), or temperature throughout the object for a given applied load or thermal condition. The stress, deformation, or temperature is often displayed as a color map on the solid model, visually displaying the regions of high stress or temperature. From these results, the engineer can redesign the object to avoid stress concentrations, large deformations, or undesirable temperatures. Likewise, the fluid flow through or around an object can be calculated by applying similar FEA techniques using the equations of fluid flow to the regions bounded by the object. These analysis tools have greatly improved and enhanced the design of many products.

Solids models can also be used in the manufacturing process. Software is available to automatically generate machine tool paths to machine the object based on the solids model. Then the software simulates the removal of material from an initial block of material on the computer. This allows the engineer to optimize the machining operation. These tool paths can be downloaded onto Computer Numerical Control (CNC) machine tools. The CNC machine tool automatically removes the desired surfaces from a block of material with high precision, allowing many identical parts to be machined automatically based directly on the solids model. Alternatively, the solids model can be downloaded onto a rapid prototyping machine. This machine automatically drives lasers that solidify liquid plastic resin, drives a robotic system to dispense a fine bead of molten plastic, or drives a laser that cuts layers of paper. In all three cases, layers are built up to become a physical model of the computer solids model, often within just a few hours. These manufacturing tools have made it possible to obtain parts from CAD models in several hours instead of several weeks.

4.3 THE NATURE OF SOLIDS MODELING

Solids modeling grew steadily in the 1980s, but it was not until Pro/ENGINEER® was introduced in 1988 and SolidWorks® was introduced in the 1990s that solids modeling delivered on its promised gains in productivity. These gains result from three characteristics of the software: feature-based, constraint-based, and associative modeling.

Feature-based modeling attempts to make the modeling process more efficient by creating and modifying geometric *features* of a solid model in a way that represents how geometries are created using common manufacturing processes. Features in a part have a direct analogy to geometries that can be manufactured or machined. A *base feature* is a solid model that is roughly the size and shape of the part that is to be modeled. The base feature is the 3-D solid created by revolving or extruding a cross-section, such as those shown in Figure 4.6. It can be thought of as the initial work block. All subsequent features reference the base feature either directly or indirectly. Additional features shape or refine the base feature. Examples of additional features include holes or cuts in the initial work block.

The analogy between feature-based modeling and common manufacturing processes is demonstrated in Figure 4.7 for making a handle for a pizza cutter. Beginning at the top of the figure we follow the steps that an engineer would use to create a solid model, or virtual part, on the left and the steps that a machinist would take to create the same physical part in a machine shop on the right. The engineer using solids modeling software begins by creating a two-dimensional profile, or *cross-section*, of a part, in this case a circle (shown in isometric projection). The analogous step by a machinist is to choose a circular bar stock of material with the correct diameter. Next the engineer *extrudes*, or stretches, the circular cross-section along the axis perpendicular to the plane of the circle to create a three-dimensional base feature (a cylinder in this case). The equivalent action by a machinist is to cut off a length of bar stock to create an initial work block. Now the engineer adds features by cutting away material on the left end to reduce the diameter and by rounding the right end of the cylinder. The machinist performs similar operations on a lathe to remove material from the cylinder. Next the engineer creates a circular cut to form a hole through the cylinder on the right end. The machinist drills a hole in the right end of the cylinder. Finally, the engineer creates a pattern of groove cuts around the handle. Likewise the machinist cuts a series of grooves using a lathe. In similar fashion a geometric shape could be added to the base feature in the solids model, analogous to a machinist welding a piece of metal to the work block. Feature-based techniques give the engineer the ability to easily create and modify common manufactured features. As a result, planning the manufacture of a part is facilitated by the correspondence between the features and the processes required to make them.

Constraint-based modeling permits the engineer or designer to incorporate "intelligence" into the design. Often this is referred to as *design intent*. Unlike traditional CAD software, the initial sketch of a two-dimensional profile in constraint-based solids modeling does not need to be created with a great deal of accuracy. It just needs to represent the basic geometry of the cross-section. The exact size and shape of the profile is defined through assigning enough parameters to fully "constrain" it. Some of this happens automatically. For example, if two nearly parallel edges are within some preset tolerance range of parallel (say 5 degrees), then the edges are automatically constrained to be parallel. As the part is resized, these edges will always remain parallel no matter what other changes are made. Likewise, if a hole is constrained to be at a certain distance from an edge, it will automatically remain at that distance from the edge, even if the edge is moved. This differs from traditional CAD where both the hole and the edge would each be fixed at a particular coordinate location. If the edge were moved, the hole location would need to be respecified so that the hole will remain the same distance from the edge. The advantage of constraint-based modeling is that the design intent of the engineer remains intact as the part is modified.

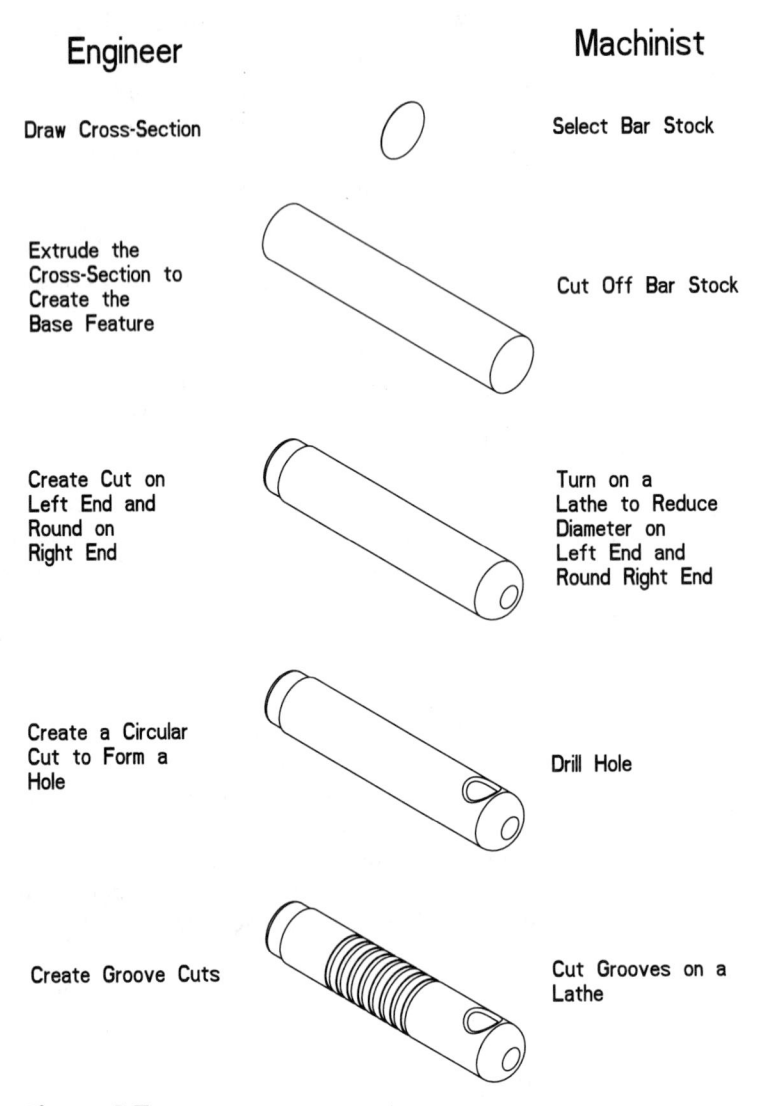

44 Chapter 4 Computer Aided Design and Drafting

Engineer		Machinist
Draw Cross-Section		Select Bar Stock
Extrude the Cross-Section to Create the Base Feature		Cut Off Bar Stock
Create Cut on Left End and Round on Right End		Turn on a Lathe to Reduce Diameter on Left End and Round Right End
Create a Circular Cut to Form a Hole		Drill Hole
Create Groove Cuts		Cut Grooves on a Lathe

Figure 4.7.

Another aspect of constraint-based modeling is that the model is *parametric*. This means that parameters of the model may be modified to change the geometry of the model. A dimension is a simple example of a parameter. When a dimension is changed, the geometry of the part is updated. Thus, the *parameter drives the geometry*. This is in contrast to other modeling systems in which the geometry is changed, say by stretching a part, and the dimension updates itself to reflect the stretched part. An additional feature of parametric modeling is that parameters can reference other parameters through relations or equations. For example, the position of the hole in Figure 4.4 could be specified with numerical values, say 40 mm from the right side of the part. Or the position of the hole could be specified parametrically, so that the center of the hole is located at a position that is one-half of the total length of the part. If the total length was specified as 80 mm, then the hole would be 40 mm from either side. However, if the length was changed to 100 mm, then the hole would automatically be positioned at half of this distance, or 50 mm from either side. Thus, no matter what the length of the part is, the hole stays in the middle of its length. The power of this approach is that when one dimension is modified, all linked dimensions are updated according to specified mathematical relations, instead of having to update all related dimensions individually.

The last aspect of constraint-based modeling is that the order in which parts are created is critical. This is known as *history-based modeling*. For example, a hole should not be created before a solid volume of material in which the hole occurs has been modeled. If the solid volume is deleted, then the hole should be deleted with it. This is known as a *parent-child relation*. The child (hole) cannot exist without the parent (solid volume) existing first. Parent-child relations are critical to maintaining design intent in a part. Most solids modeling software recognizes that if you delete a feature with a hole in it, you do not want the hole to remain floating around without being attached to a feature. Consequently, careful thought and planning of the base feature and initial additional features can have a significant effect on the ease of adding subsequent features and making modifications.

The *associative* character of solids modeling software causes modifications in one object to "ripple through" all associated objects. For instance, suppose that you change the diameter of a hole on the engineering drawing that was created based on your original solid model. The diameter of the hole will be automatically changed in the solid model of the part, too. In addition, the diameter of the hole will be updated on any assembly that includes that part. Similarly, changing the dimension in the part model will automatically result in updated values of that dimension in the drawing or assembly incorporating the part. This aspect of solids modeling software makes the modification of parts much easier and less prone to error.

As a result of being feature-based, constraint-based, and associative, solids modeling captures "design intent," not just the design. This comes about because solids modeling software incorporates engineering knowledge into the solid model with features, constraints, and relationships that preserve the intended geometrical relationships in the model.

PROFESSIONAL SUCCESS

Has CAD Impacted Design?

CAD has not only automated and provided a more accurate means of creating engineering drawings, it has created a new paradigm for design. Parts can be modeled, visualized, revised, and improved on the computer screen before any engineering drawings have even been created. Parts that have been modeled can be assembled in the virtual environment of the computer. The relative motion of moving parts can be animated on the computer. The stresses in the parts can be assessed computationally and the part redesigned to minimize stress concentrations. The flow of fluids through the part or around the part can be modeled computationally. The machine tool path or mold-filling flow to fabricate the part can be modeled on the computer. The part model can be downloaded to a rapid prototyping system that can create a physical model of the part in a few hours with virtually no human intervention.

Although some of the analyses described above are outside of the capability of the CAD software itself, the ability to produce a CAD model makes these analyses possible. Without the ability to create a complex solid geometry computationally using CAD, much of the engineering analysis that is used to design, improve, and evaluate a new design would be much more difficult.

KEY TERMS

Associative modeling
Base Feature
Constraint-based modeling
Design intent
Extrude
Feature
Feature-based modeling
History-based modeling
Parametric modeling
Parent-child relation
Revolve
Solids modeling
Surface model
Wireframe model

46 Chapter 4 Computer Aided Design and Drafting

Problems

1. The wireframe model shown in Figure 4.8 is a small cube centered within a large cube. The nearest corners of the two cubes are connected. For instance, the lower left front corners of both cubes are connected with a line. This wireframe model is ambiguous; that is, it could represent several different solid bodies. Freehand sketch or trace the figure. Then use dashed lines to indicate hidden lines to show two different solids model configurations that are possible for this wireframe model.

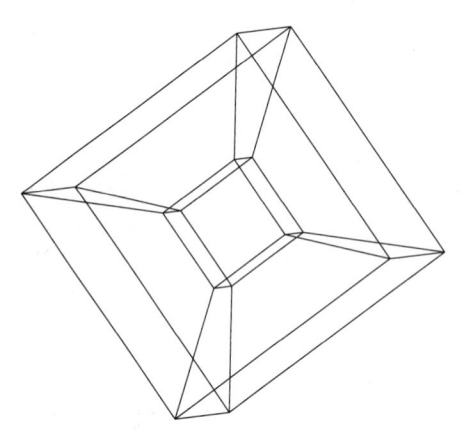

Figure 4.8.

2. Using a technique similar to that shown in Figure 4.5, freehand sketch the geometric primitives that could be added or subtracted to form the objects shown in Figure 4.9. (The hole in Figure 4.9b is only through the face shown, not the back face. The holes in Figures 4.9c and 4.9d are through holes.)

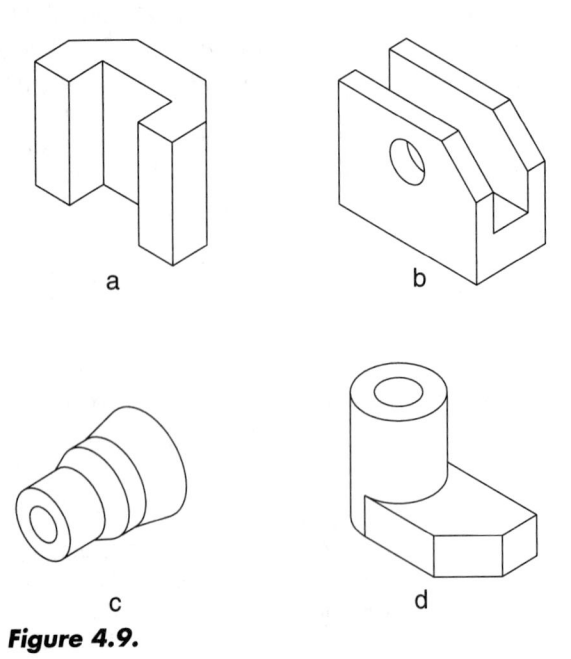

a

b

c

d

Figure 4.9.

3. Sketch the cross-section that should be revolved to create the objects shown in Figure 4.10. Include the axis about which the cross-section is revolved in the sketch.

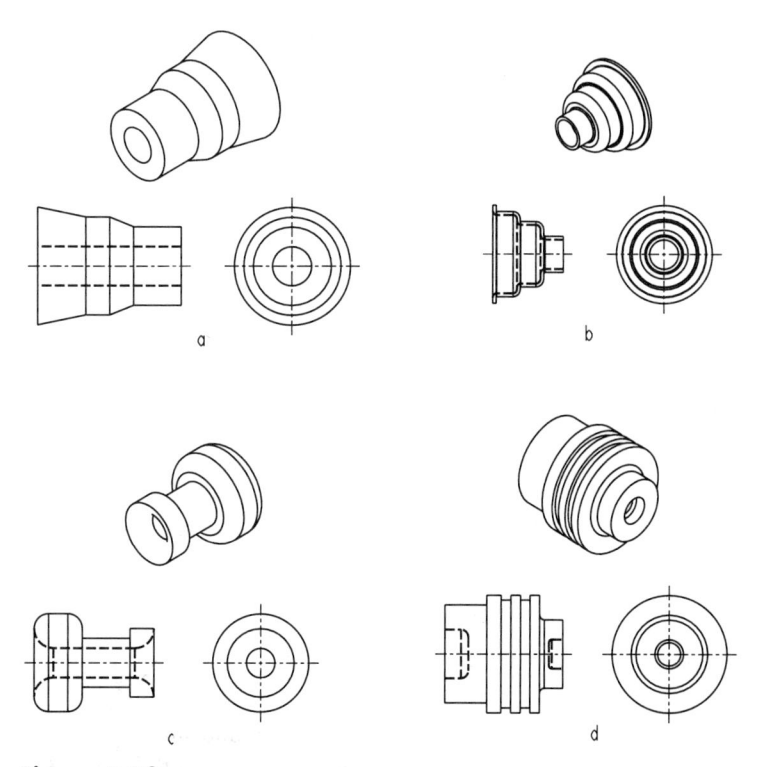

Figure 4.10.

4. Sketch at least two cross-sections that could be extruded to form a base feature for the objects shown in Figure 4.11. Use cross-sections that can be extruded into a base feature from which material is only removed, not added. Make sure that the cross-sections that you propose minimize the number of additional features necessary to modify the base feature. (The hole in Figure 4.11a is only through the face shown, not the back face. The holes in Figure 4.11b, 4.11c, and 4.11e are through holes.)

5. Consider the following objects. Sketch what the base feature would look like. List what features would be added to model the object. The type of base feature to be used (extrude, revolve, or both) is noted.

    a. hexagonal cross-section wooden pencil that is sharpened to a point (do not include eraser)—extrude.
    b. plastic 35 mm film container and cap—extrude and revolve.
    c. nail—extrude and revolve.
    d. push pin—revolve.
    e. baseball bat—revolve.
    f. broom handle—revolve and extrude.
    g. ceiling fan blade—extrude.

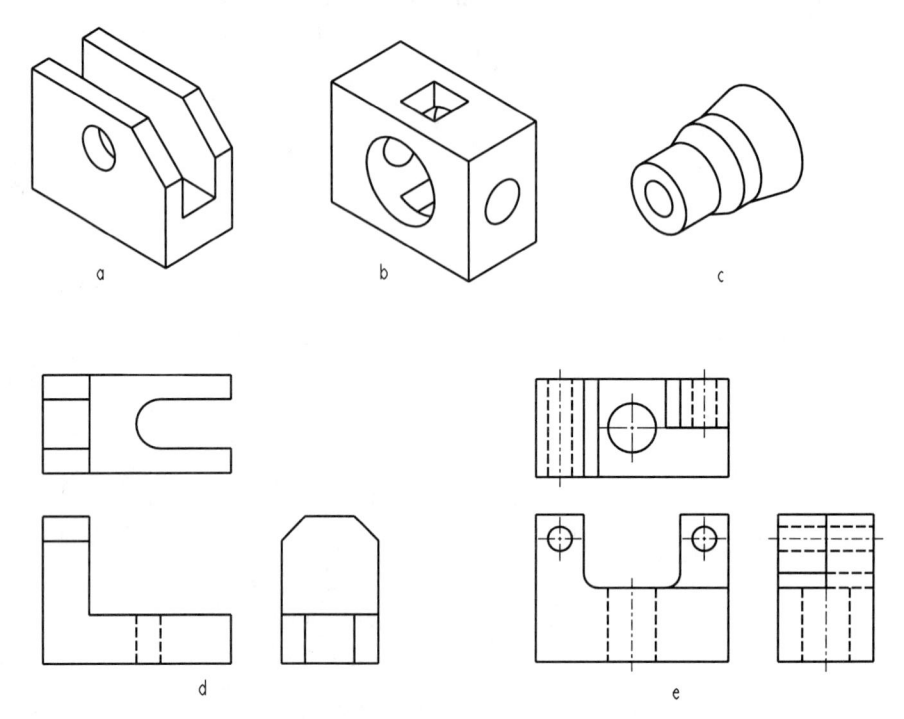

Figure 4.11.

      h.   cinder block—extrude.
      i.   compact disk—revolve and extrude.
      j.   single staple (before being deformed)—extrude.
      k.   automobile tire—revolve.
      l.   gear—extrude.
      m.   round toothpick—revolve.

6. Describe parametric modeling and give an example of a family of parts where parametric modeling would be useful.
7. Describe constraint-based modeling and explain how it relates to design intent.

5

Standard Practice for Engineering Drawings

Overview

Drawing standards and conventions are used to clarify engineering drawings and simplify their creation. For example, standard sizes for drawings are used, and standard types and weights of lines designate different items on a drawing. The proper placement of dimensions on drawings is helpful in making the drawing more readable. In addition, internal or small details of the part can be displayed using special views. Screw threads are difficult to draw, so standard representations are used for screw threads.

Sections

- 5.1 Introduction to Drawing Standards
- 5.2 Sheet Layouts
- 5.3 Lines
- 5.4 Dimension Placement and Conventions
- 5.5 Section and Detail Views
- 5.6 Fasteners and Screw Threads
- 5.7 Assembly Drawings

Objectives

After reading this chapter, you should be able to:

- Explain why drawing standards are used
- Choose the proper sheet layout
- Read and understand the scale of a drawing
- Differentiate linetypes used in engineering drawings
- Properly place dimensions on a drawing
- Read and understand section and detail views on a drawing
- Read and understand screw thread designations on a drawing
- Read an assembly drawing

5.1 INTRODUCTION TO DRAWING STANDARDS

Not only is the accurate and clear depiction of details of a part necessary in an engineering drawing, but it is also necessary that the drawing conforms to commonly accepted standards, or conventions. There are two reasons for the existence of standards and conventions. First, using standard symbols and projections ensures clear interpretation of the drawing by the viewer. An example of the problem of differing presentations is the use of the third-angle orthographic projection in North America and the first-angle orthographic projection elsewhere in the world. (Recall that the top and bottom views and the right and left views are reversed in third-angle and first-angle projections.) This can bring about confusion if, for example, a U. S. engineer tries to interpret a German drawing. If a single standard existed in the world, there would be no confusion. A second reason for using conventions is to simplify the task of creating engineering drawings. For example, the symbol Ø associated with a dimension indicates that the dimension is a diameter of a circular feature. Without this convention, it would be more difficult to unambiguously represent diameter dimensions on a drawing.

Most CAD programs automatically use a standardized presentation of drawings, usually based on standards from the American National Standards Institute (ANSI) in the United States and the International Standards Organization (ISO) in the rest of the world. Even with the use of CAD software to assure standard presentation in drawings, it is up to the engineer to implement drawing standards in some cases. An example is in the placement of dimensions in a drawing. The user controls where dimensions appear on the drawing. Even in cases where the CAD software automatically implements drawing standards, it is still necessary to understand what the conventions mean so they are properly interpreted.

5.2 SHEET LAYOUTS

Engineering drawings are created on standard size sheets that are designated by the code indicated in Table 5.1. The ISO drawing sizes are just slightly smaller than ANSI sizes. Engineering drawings are almost always done in "landscape" orientation so that the long side of the drawing is horizontal. The choice of drawing size depends upon the complexity of the object depicted in the drawing. The drawing should be sized so that the projections of the part, dimensions, and notes all fall within the borders of the drawing with adequate spacing so that the drawing is not cluttered. The drawing should be large enough so that all details are readily evident and readable. As a result, regardless of the physical size of the part, simple parts are usually drawn on smaller sheets because it is not necessary to show much detail. Complex parts are usually drawn on larger sheets to readily show adequate detail. When using CAD software, it is necessary to consider the printed drawing size in addition to the drawing size on the computer screen. On the computer screen, it is possible to Zoom In or Zoom Out to read detail, but this cannot be done for a printed drawing. Consequently, the sheet size should be chosen carefully.

TABLE 5-1 Standard Sheet Sizes

ANSI	ISO
A—8.50″ × 11.00″	A4—210 mm × 297 mm
B—11.00″ × 17.00″	A3—297 mm × 420 mm
C—17.00″ × 22.00″	A2—420 mm × 594 mm
D—22.00″ × 34.00″	A1—594 mm × 841 mm
E—34.00″ × 44.00″	A0—841 mm × 1189 mm

The *title block* on a drawing records important information about the working drawing. Normally the title block of a drawing is in the lower right corner of the drawing. ANSI standard title blocks can be used, but often individual companies use their own standard title block. The title block, such as the one shown in Figure 5.1, is used primarily for drawing control within a company. The title block includes information regarding the part depicted in the drawing, such as its name and part number; the person who created the drawing; persons who checked or approved the drawing; dates the drawing was created, revised, checked, and approved; the drawing number; and the name of the company. This information allows the company to track the drawing within the company if questions arise about the design.

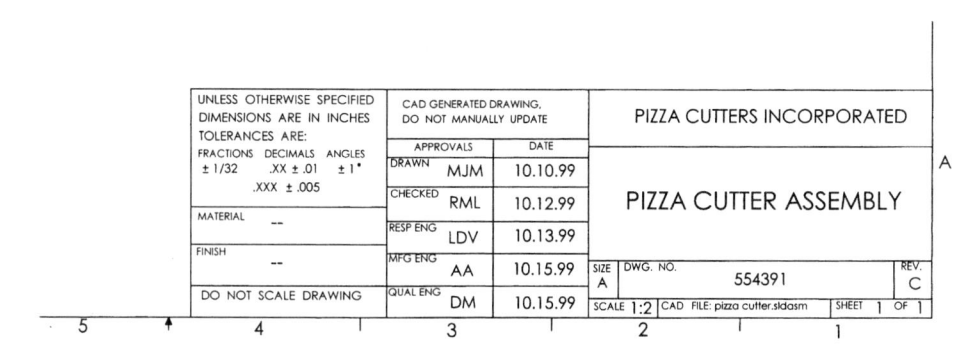

Figure 5.1.

Other information can appear in the title block to provide details about the manufacture of the part, such as the material for the part, the general tolerances, the surface finish specifications, and general instructions about manufacturing the part (such as "remove all burrs"). Many times, though, this information is shown as notes on the drawing rather than in the title block.

Finally the title block provides information about the drawing itself, such as the units used in dimensioning (typically inches or millimeters) and the scale of the drawing. The *scale* of the drawing indicates the ratio of the size on the drawing to the size of the actual object. It can be thought of as the degree to which the object is enlarged or shrunk in the drawing. Scales on drawings are denoted in several ways. For instance, consider an object that is represented in the drawing as half of its actual size. Then the drawing is created such that 1 inch on the drawing represents 2 inches on the physical object. This can be reported in the Scale box in the title block as HALF SCALE, HALF SIZE, 1 = 2, 1:2, or 1/2″ = 1″. In the numerical representations, the left number is the length on the drawing and the right number is the length of the actual object. A part that is drawn twice as large as its physical size would be denoted as DOUBLE, 2× (denoting "2 times"), 2 = 1, or 2:1. In some cases, such as architectural drawings, the scale can be quite large. For instance 1/4″ = 1′-0″ corresponds to a scale of 1:48, where ″ denotes inches and ′ denotes feet. Common scales used in engineering drawing are 1 = 1, 1 = 2, 1 = 4, 1 = 8, and 1 = 10 for English units and 1:1, 1:2, 1:5, 1:10, 1:20, 1:50, and 1:100 for metric units. A 1:1 scale is often specified as FULL SCALE or FULL SIZE.

5.3 LINES

The types of lines used in engineering drawing are standardized. Different linetypes have different meanings and a knowledge of linetypes makes interpreting drawings easier. The linetypes vary in thickness and in the length and number of dashes in dashed lines. Most CAD software automatically produces linetypes that correspond correctly to the application. Figure 5.2 shows some of the more commonly used linetypes, each of which is described below:

Visible lines (thick, solid lines) represent the outline of the object that can be seen in the current view.

Hidden lines (thin, dashed lines) represent features that are hidden behind surfaces in the current view.

Centerlines or *symmetry lines* (thin, long-dash/short-dash lines) mark axes of rotationally symmetric parts or features.

Dimension lines (thin lines with arrowheads at each end) indicate sizes in the drawing.

Extension lines or *witness lines* (thin lines) extend from the object to the dimension line to indicate which feature is associated with the dimension.

Leader lines (thin, solid lines terminated with arrowheads) are used to indicate a feature with which a dimension or note is associated.

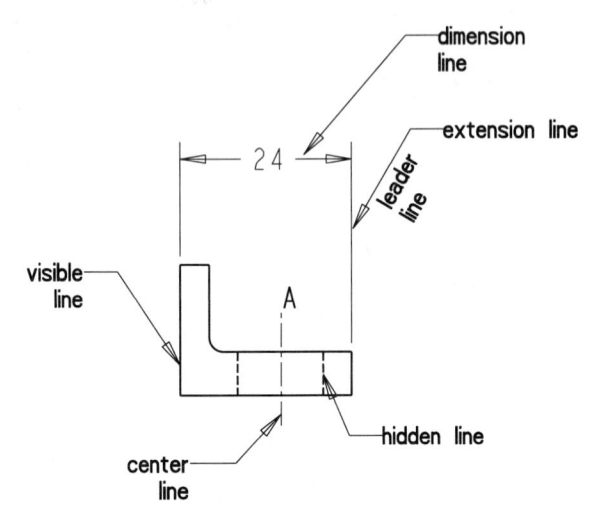

Figure 5.2.

Other linetypes are used for break lines (lines with zigzags) that show where an object is "broken" in the drawing to save drawing space or reveal interior features and cutting plane lines (thick lines with double short dashes and perpendicular arrows at each end) that show the location of cutting planes for section views.

The arrowheads for dimension lines, leader lines, and other types of lines may be filled or not filled. Most CAD software uses unfilled arrowheads as a default, although many engineering graphics textbooks prefer filled arrowheads.

It is common in technical drawings that two lines in a particular view coincide. When this happens, a convention called the *precedence of lines* dictates which line is shown. Visible lines have precedence over hidden lines, which have precedence over centerlines, which have precedence over extension lines.

Conventions also exist for the intersection of lines in a drawing. For instance, an extension line is always drawn so that there is a slight, visible gap between its end and the outline of the object as shown in Figure 5.2. Similarly, perpendicular centerlines that cross are drawn so that the short dashes of each line intersect to form a small cross. Fortunately, most CAD software takes care of all of the details related to linetypes automatically. The engineer or designer need only know how to properly interpret the various linetypes.

5.4 DIMENSION PLACEMENT AND CONVENTIONS

Engineering drawings are usually drawn to scale, but it is still necessary to specify numerical dimensions for convenience and to ensure accuracy. In general, sufficient dimensions must be provided to define precisely the geometry of the part, but redundant dimensions should be avoided. Furthermore, dimensions on the drawing should reflect the way the part is made or the critical dimensions of the part. For example, consider the three drawings of a plate with two holes shown in Figure 5.3. The left drawing is over-dimensioned, having one redundant dimension. It is clear that given the distances from the left edge to the left hole, the distance from the right edge to the right hole, and the overall width of the plate, the distance between the holes has to be 6-mm. Thus, the 6-mm dimension could be omitted, suggesting that the distance between each hole and the nearest edge of the plate is most important. But suppose that the distance between the holes is critical. This might be necessary if the holes in the plate are supposed to align with another part. Then the dimensions on the middle drawing would clearly show that the distance between the holes is most important by omitting the dimension from the right hole to the right edge. The right drawing indicates that the distance from the left edge to each hole, not the distance between the holes, is critical. The engineer or designer must consider the optimal way to display dimensions to clearly show which dimensions are most important.

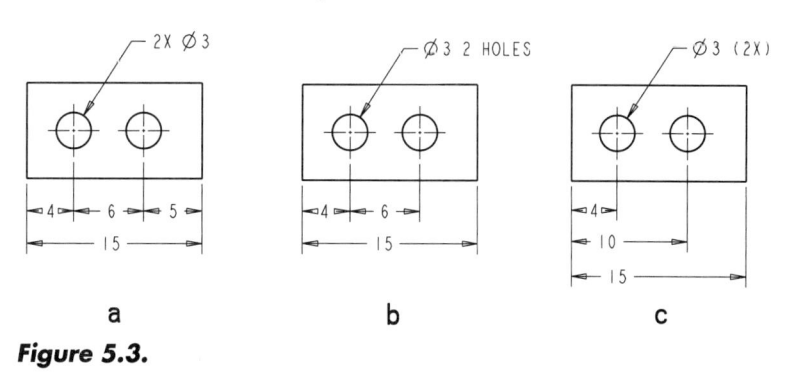

Figure 5.3.

It is natural for a person reading the drawings to look for the dimensions of a feature in the view where the feature occurs in its most characteristic shape and where it is visible (as opposed to hidden). This is known as *contour dimensioning*. For example, the location and size of a hole should be dimensioned in the view where the hole appears as a circle. Likewise, it is best to show the overall dimensions of the object in a view that is most descriptive of the object. If possible both the horizontal and vertical location of a feature should be dimensioned in the same view. The upper part of Figure 5.4 shows an object properly dimensioned in millimeters to show the relation of the dimensions to the features in the most obvious manner. In the lower part of Figure 5.4, the dimensioning has several problems. The overall dimensions of the L-shaped profile are shown on the view where it is not clear that the object is L-shaped. The diameter of the hole is shown in a view where the hole is shown only as a pair of parallel hidden lines, not as a circle. And the vertical location of the hole is shown in one view, while the horizontal location is shown in another view.

The conventions for millimeter and inch dimensioning are different. For millimeter dimensioning, dimensions less than 1 mm have a zero preceding the decimal point, and integer dimensions may or may not include a decimal point and following zeros. Thus, legitimate millimeter dimensions are 0.8, 6, 8.00, and 1.5. Inch dimensions do not include a zero before the decimal point for values less than 1 inch. Zeros to the right of the decimal point are usually included. Legitimate inch dimensions are .800, 6.00, 8.00, and 1.500. For both millimeter and inch dimensions, the number of digits to the right of the decimal point may be used to indicate the tolerance (acceptable variation). It is standard practice to omit the units for dimensions. However, it is wise to note the units in the title block or as a note on the drawing.

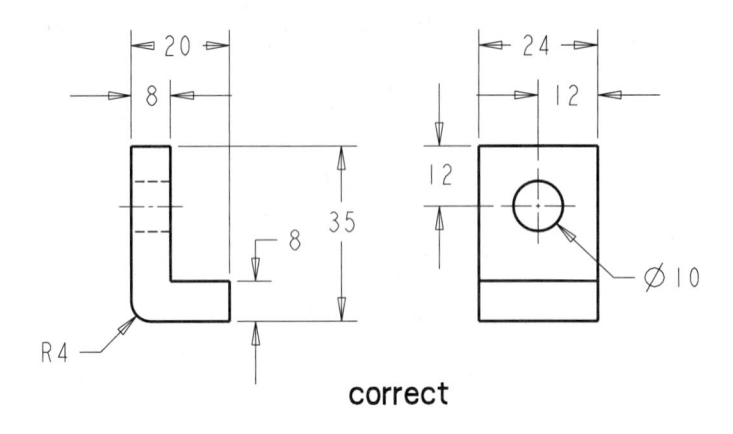

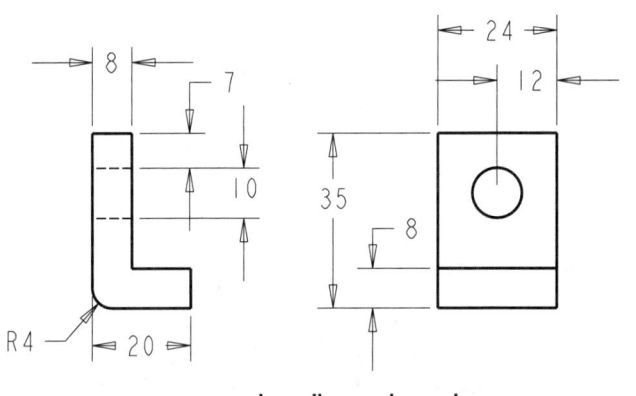

Figure 5.4.

Circles and arcs are dimensioned using special symbols and rules. A radius is denoted using a leader line with an arrow pointing at the arc as shown in Figure 5.4. Sometimes a small cross is placed at the center of the radius. The position of the center should be dimensioned unless the radius is for a rounded edge, in which case the arc is positioned by virtue of being tangent at its ends to the sides forming the corner that is rounded, as shown in Figure 5.4. The dimension for a radius begins with a capital letter R to indicate radius. Full circles are dimensioned using the diameter with a leader line having an arrow pointing at the circle, as shown in Figure 5.4. If space permits, the leader line extends across a diameter of the circle with arrows at each end of the diameter. The dimension for a diameter begins with the Greek letter phi (Ø) to indicate diameter. The location of the center of the circle must also be dimensioned. Sometimes maintaining clarity makes it necessary to dimension several concentric circles from a side view, where the circle does not appear as a circle, thus violating the rule to dimension a feature in its most characteristic shape.

Angles are dimensioned as decimals or in degrees and minutes (60 minutes is one degree), so that 32.5° and 32°30′ are equivalent, where ′ indicates minutes. The dimension line for an angle is drawn as an arc with its center at the apex of the angle. When lines are drawn at right angles, a 90° angle is implied.

Multiple features that are identical need not be individually dimensioned. For example, the diameter of the two 3-mm holes in Figure 5.3 could be dimensioned with a leader line to one hole with the dimension 2× Ø3. The 2× indicates "two times." Alternative dimensions are Ø3 (2×), Ø3 2 PLACES, or Ø3 2 HOLES.

There are several other "rules" for dimensioning:

- Dimension lines should be outside of the outline of a part whenever possible.
- Dimension lines should not cross one another.
- Dimensions should be indicated on a view that shows the true length of the feature. This is particularly important when dimensioning features on a surface that is at an angle to the plane of the orthographic view.
- Each feature should be dimensioned only once. Do not duplicate the same dimension in different views.
- Dimension lines should be aligned and grouped when possible to promote clarity and uniform appearance. See for example the placement of the horizontal dimensions in Figure 5.3c.
- The numerical dimension and arrows should be placed between the extension lines where space permits. If there is space only for the dimension, but not the arrows, place the arrows outside of the extension lines. When the space is too small for either the arrows or the numerical value, place both outside of the extension lines. See Figure 5.4 for examples.
- Place the dimension no closer than about 10 mm (3/8 inch) from the object's outline.
- Dimensions should be placed in clear spaces, as close as possible to the feature they describe.
- When dimensions are nested (such as the horizontal dimensions in Figure 5.3c) the smallest dimensions should be closest to the object.
- Avoid crowding dimensions. Leave at least 6 mm (1/4 inch) between parallel dimension lines.
- Extension lines may cross visible lines of the object.
- Dimensions that apply to two adjacent views should be placed between the views, unless clarity is enhanced by placing them elsewhere. The dimension should be attached to only one view. Extension lines should not connect two views.
- Dimension lines and extension lines should not cross, if possible. Extension lines may cross other extension lines.
- A centerline may be extended to serve as an extension line, in which case it is still drawn as a centerline.
- Centerlines should not extend from view to view.
- Leader lines are usually sloped at about 30°, 45°, or 60° and are never horizontal or vertical.
- Numerical values for dimensions should be centered between arrowheads, except in a stack of dimensions (like Figure 5.3c).
- When a rough, noncritical dimension (such as a round) is indicated on a drawing, add the note TYP to the dimension to indicate that the dimension is "typical" or approximate.

5.5 SECTION AND DETAIL VIEWS

The cutaway view of a device appeared first in various forms in the 15th and 16th centuries to show details of parts hidden by other elements. These cutaway views have evolved to *section views* in which interior features that cannot be effectively displayed by hidden lines are exposed by slicing through a section of the object. To create a section view, a cutting plane is passed through the part and the portion of the part on one side of the cutting plane is imagined to be removed. In a section view, all visible edges and contours behind the cutting plane are shown. Hidden lines are usually omitted. The portion of the object that is sliced through is designated with angled crosshatch lines known as section-lining.

Key to the interpretation of a section view is the clear representation of where the cutting plane is and from which direction it is viewed. This is accomplished using a *cutting plane line* (long dash - short dash - short dash - long dash with perpendicular arrows at each end). The cutting plane line shows where the cutting plane passes through the object. The arrows on either end of the cutting plane line show the direction of the line of sight for the section view. An example of a block with a hole that has a large diameter partway into it and a smaller diameter through it is shown in Figure 5.5. In this case, the object is sectioned along a plane parallel to the right and left edges of the front view and through the center of the hole, as indicated by the position of cutting plane line A-A. The arrows indicate that the section view will be shown as if we are looking in the direction of the arrows. Thus, the section view shown to the right of the front view (as if unfolding the glass box) shows what the viewer would see if the portion of the object to the right of the cutting plane line was removed. The viewer sees the two diameters of the hole with the material that has been "cut" shown as cross-hatched. This view is designated Section A-A to correspond to the cutting plane line A-A. This type of section in which a single plane goes completely through an object is known as a *full section*.

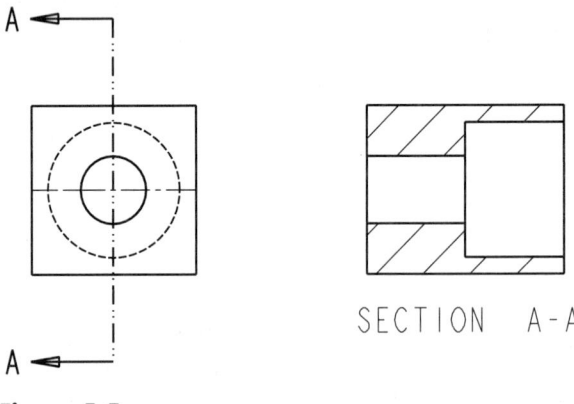

Figure 5.5.

A slightly more complicated example of a full section is shown in Figure 5.6. The whole object is shown in the upper left portion of the figure, and the sectioned object is shown with a portion removed just below it. The right side of the figure shows a top view of the object with the cutting plane displayed. Imagine that the material on the side of the cutting plane in the direction the arrows are pointing is retained, and the material behind the arrows is removed. Then the projection of the retained portion appears as SECTION B-B when viewed in the direction of the arrows.

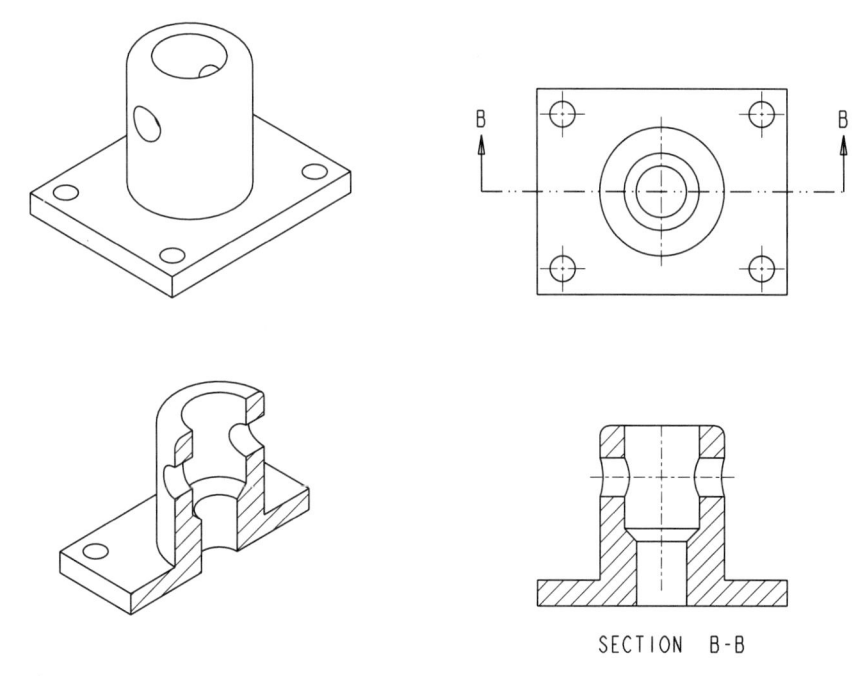

Figure 5.6.

Much more sophisticated section views can be created for complicated parts. In all cases, though, the clear definition of the cutting plane line and the direction of the line of sight for the view are critical. For example, Figure 5.7 shows a *half section* of the same part. The cutting plane extends halfway through the object to show the interior of one half of the object and the exterior of the other half. This type of section is ideal for

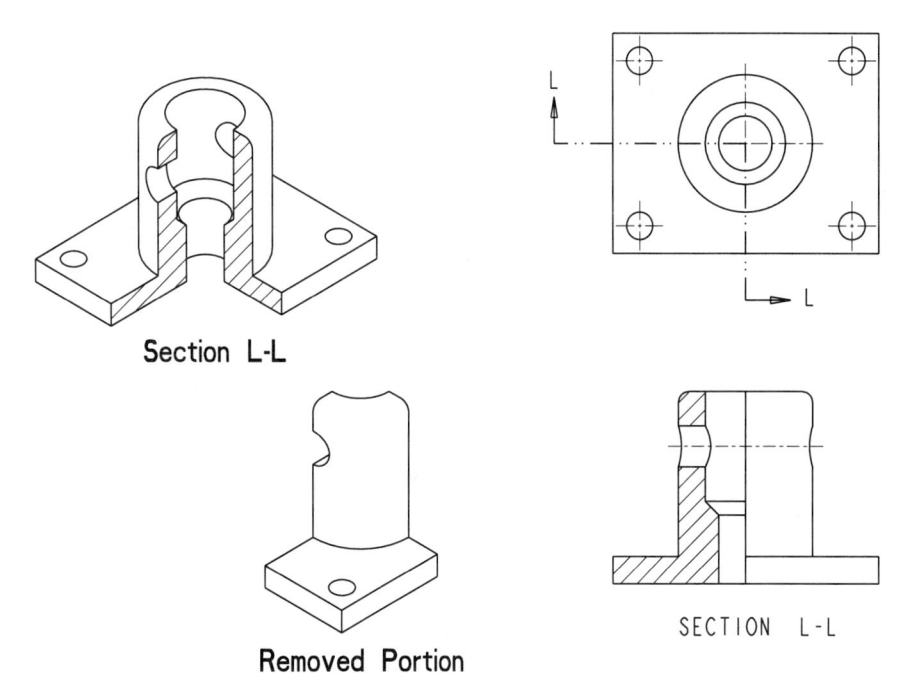

Section L-L

Removed Portion

SECTION L-L

Figure 5.7.

symmetrical parts in which it is desired to show internal and external features in a single view. An *offset section* like that shown in Figure 5.8 is used to show internal details that are not in the same plane. The cutting plane for offset sections is bent at 90° angles to pass through important features. Note that the change of plane that occurs at the

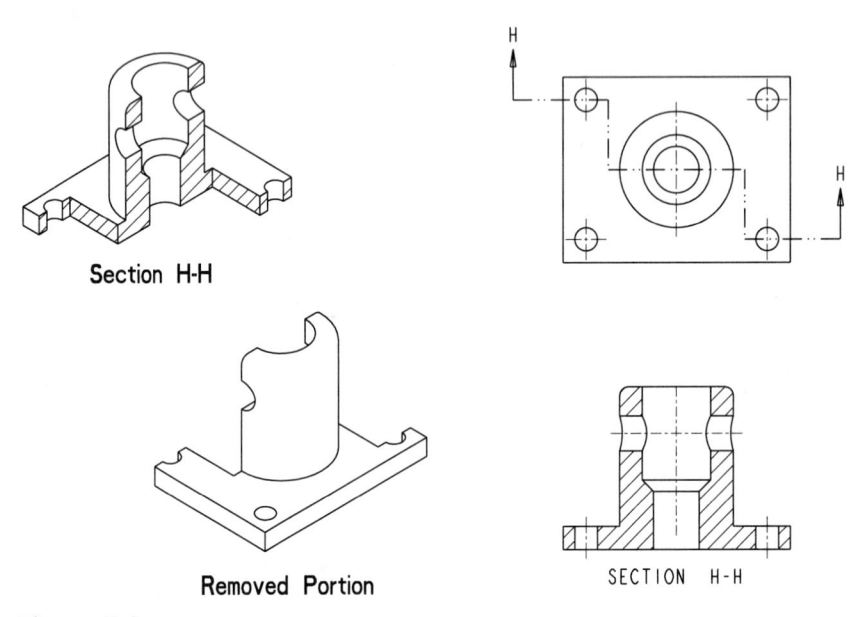

Figure 5.8.

90° bends is not represented with lines on the section view. When only a portion of the object needs to be sectioned to show a particular internal detail, a *broken-out section* like that shown in Figure 5.9 can be used. A break line separates the sectioned portion from the unsectioned portion, but no cutting plane line is drawn.

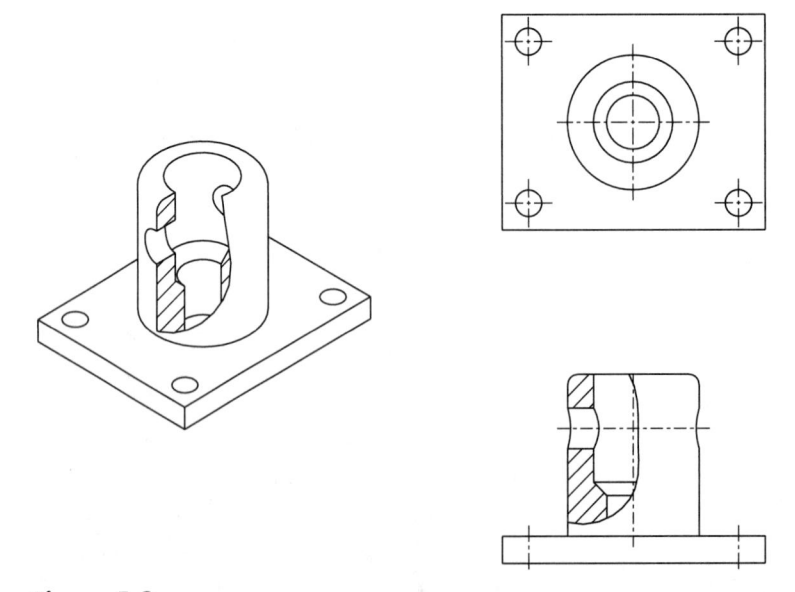

Figure 5.9.

In some cases it is necessary to clarify details of the part that may not be readily evident in the views normally shown in the drawing. For instance, a particular detail may be so small that it cannot be shown clearly on the same scale as the remainder of the part. This is done using an auxiliary *detail view* consisting of a small portion of the object magnified to make clear the small feature. An example is shown in Figure 5.10. In this case the small rivet holding the blade onto the arms of the pizza cutter is too small to see in the orthographic views. The region of interest near the rivet is circled in the left-most projection of the pizza cutter, and a note directs the reader to a magnified detail view in the upper right part of the drawing. The cross section of the rivet is clearly visible in this detail view.

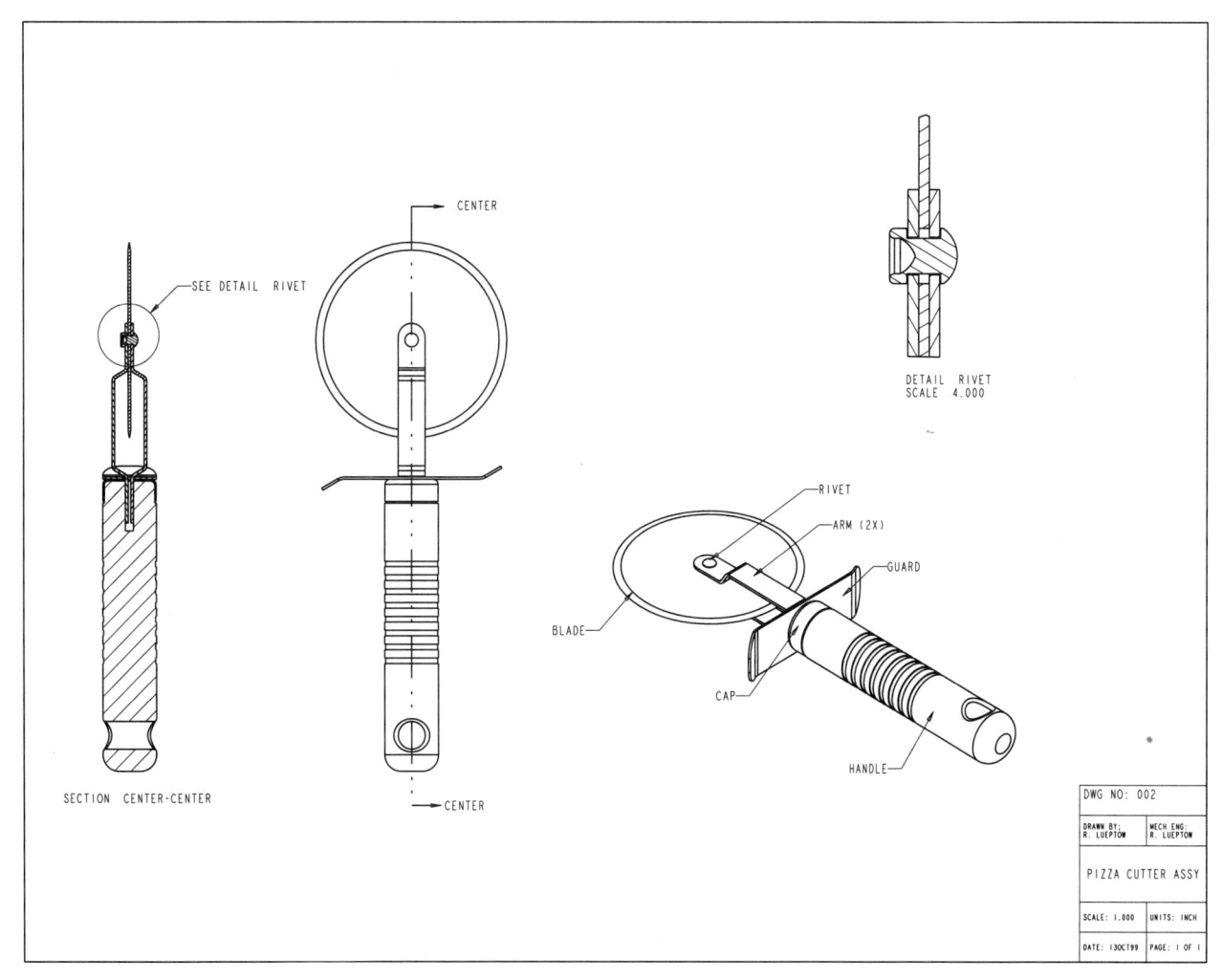

Figure 5.10.

5.6 FASTENERS AND SCREW THREADS

Fasteners include a broad range of items such as bolts, nuts, screws, and rivets used to "fasten" parts together. In most cases, fasteners are standard parts purchased from an outside vendor, so detail drawings of fasteners are rarely necessary. Nevertheless, threaded holes and threaded shafts must be represented on a drawing.

The geometry of screw threads is too complicated to draw exactly in an engineering drawing and screw threads are standard, so either of two simple conventions is used to indicate screw threads, as shown in Figure 5.11. The *schematic representation* is used when a realistic representation of the side view of a screw thread is desired. For an external thread, the lines that extend across the entire diameter represent the *crest*, or peak, of the thread, while the shorter lines in between represent the *root*, or valley, of the thread. The distance between crests or between roots is called the *pitch*. On a drawing, the crests and roots are shown perpendicular to the axis of the threaded section rather than helical as they actually appear on the physical thread. In the end view, threads are depicted with concentric circles. The outer circle is the largest diameter of the screw thread, known as the *major diameter*, and the inner circle is the smallest diameter of the screw thread, or *minor diameter*. In the end view of the external thread, both diameters are shown as solid circles. In the *simplified representation* of external screw threads, the threads are omitted altogether. The major diameter is represented as a solid line in the side view, and the minor diameter is represented with a dashed line.

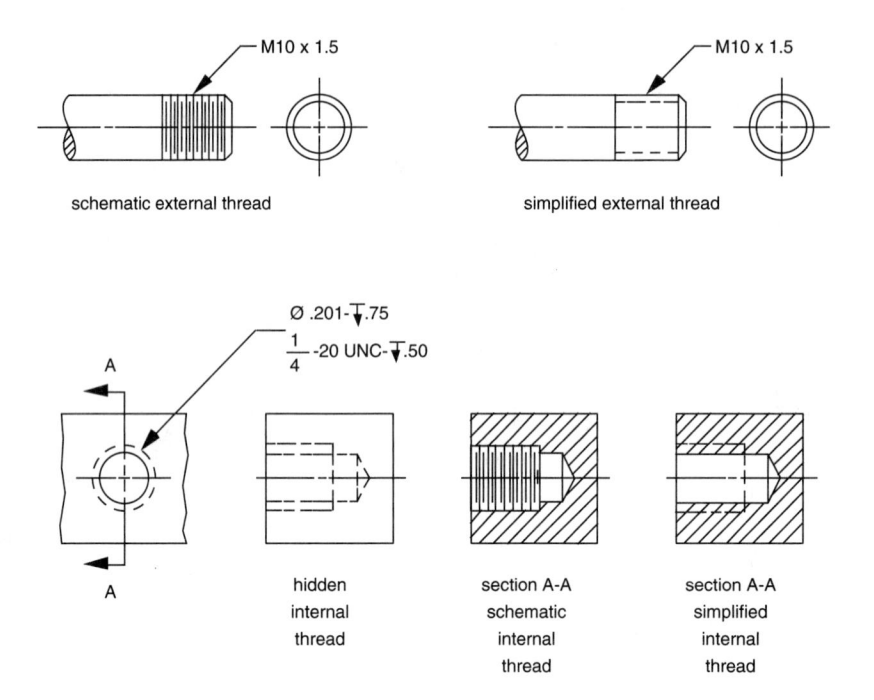

Figure 5.11.

The end view of an internal thread is shown as two concentric circles representing the major and minor diameter, but the major diameter is a hidden line. For a side view of a hidden internal screw thread, the major and minor diameters are both shown as hidden lines for both schematic and simplified screw thread representations. In a cross-section, internal threads can be shown in either the schematic or the simplified representation, as shown. For a hole that does not go completely through the part, called a *blind hole*, the lines in the side view representing the minor diameter continue deeper than the thread and come to a point. This represents the hole that is drilled prior to tapping, or cutting threads in the hole. The hole needs to be longer than the threaded section to permit the tool that is used to cut the threads, or tap, to penetrate deeply enough to fully cut the threads in the portion of the hole to be tapped.

Screw threads are specified in terms of the nominal (major) diameter, the pitch, and the thread series. For instance, the designations 1/4-20 UNC or .25-20 UNC or 1/4-20 NC all indicate a major diameter of .25 inches, a pitch of 20 threads per inch, and the United Course series. The diameter of the hole drilled before tapping an internal thread is specified in a wide variety of machinist and engineering handbooks. In this case the proper hole to be drilled has a diameter of .2010 inches, which corresponds to a number 7 drill. To make an internal thread of this size, a machinist would first drill a hole using a number 7 drill bit and then cut the threads using a 1/4-20 tap. For screw threads with a nominal diameter less than .25 inches, a number designation is used to specify the nominal diameter. For example, the designation 10-32 UNF indicates a nominal diameter of .1900 inches, 32 threads per inch, and the United Fine series. The pitch is equal to 1 divided by the number of threads per inch.

Metric threads are specified in a slightly different way. A designation M10 × 1.5 indicates a metric thread with a 10 mm major diameter and a pitch of 1.5 mm between crests of the thread. The number of threads per mm is 1 divided by the pitch. Other letters and numbers may follow the thread specification to denote tolerances and deviations of the thread, but for many cases these are not necessary. In addition, for a blind hole, the depth of the thread can be specified in several ways. Following the thread designation, the notation X .50 DEEP, THD .50 DP, or a downward arrow with a horizontal bar at its tail followed by .50 all indicate that the thread should extend .50 inches below the surface of the piece, as shown in Figure 5.11. (THD indicates "thread," and DP indicates "deep.") The threaded depth is usually 1.5 to 2 times the nominal diameter of the screw thread.

The dimensions for a threaded hole are indicated on a drawing using a leader line with the arrow pointing at the major diameter, as shown in Figure 5.11. Threaded holes are usually dimensioned in the view where they appear as circles, rather than in a side view of the threaded hole. Multiple threaded holes of the same specification are typically denoted in the same way as other multiple features using the notation (2X) or 2 TIMES at the end of the thread designation.

Nuts and bolts are usually only shown in a parts list and not on any drawings. They are specified in the same way as threaded holes, plus a bolt length and the type of head. For instance .25-28 UNF X 1.50 HEXAGON CAP SCREW or .25-28 NF X 1.50 HEX CAP SCR indicates a bolt that is 1.5 inches long with hexagonal head. Nuts are specified in terms of their thread and shape, such as M8 X 1.25 HEX NUT.

5.7 ASSEMBLY DRAWINGS

An assembly drawing shows all of the components of a design either assembled or in an exploded view. Many times assembly drawings include sections. Most dimensions are omitted in assembly drawings. Individual parts are not dimensioned, but some dimensions of the assembled mechanism may be included. Hidden lines are seldom necessary in assembly drawings, although they can be used where they clarify the design. Leader lines attached to a ballooned letter or detail number, as shown in Figure 5.12, reference the parts of the assembly. The leader lines should not cross and nearby leader lines should be approximately parallel. Sometimes parts are labeled by name rather than number. The parts list may be on the assembly drawing (usually on the right side or at the bottom) or it may be a separate sheet. The assembly drawing may also include machining or assembly information in the form of notes on the drawing.

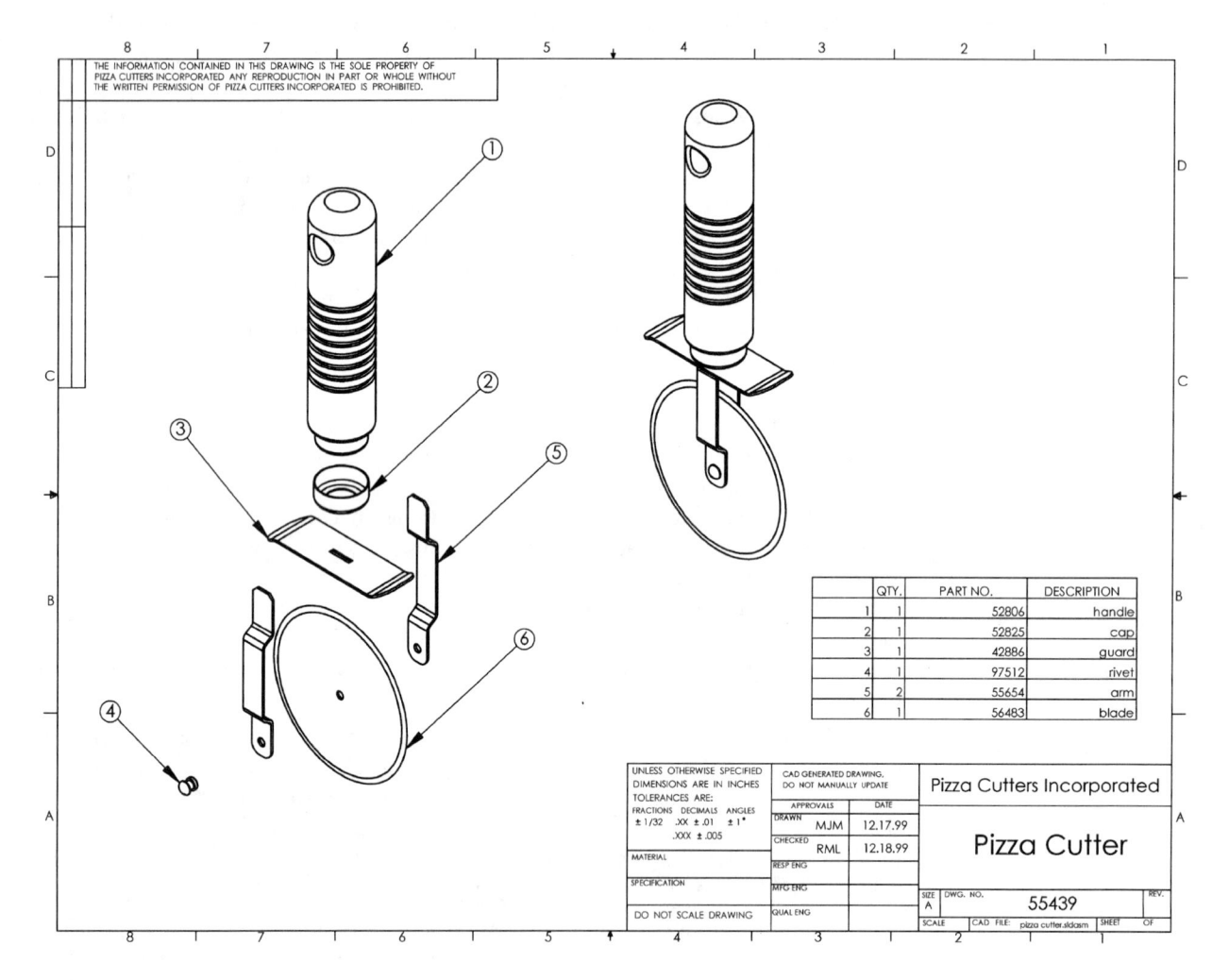

Figure 5.12.

Often assembly drawings include assembly sections. These are typically orthographic or pictorial section views of parts as put together in an assembly. Adjacent parts in assembly drawings are cross-hatched at different angles to make the separate parts clear. The assembly cross-section on the left side of Figure 5.10 shows the interior structure of the parts of a pizza cutter and how they fit together. The detail view in the upper right of Figure 5.10 shows the cross-section through the rivet to indicate how the parts are assembled. Usually standard parts such as fasteners, washers, springs, bearings, and gears are not cross-hatched. But in the case of Figure 5.10, the rivet detail is integral to the assembly, and it is helpful to show its cross-section with cross-hatching to indicate how it fits with the other parts.

KEY TERMS

Centerlines
Contour dimensioning
Cutting plane lines
Detail view
Dimension lines
Extension lines
Hidden lines
Leader lines
Scale
Section views
Title block
Visible lines

Problems

1. Three orthographic views and an isometric view of an object that has overall dimensions of 4 × 4 × 4 inches are to be shown in a drawing. Determine the best ANSI sheet size for an object of this size at the scales indicated below:

   a. Half Size.

   b. 2:1.

   c. 1 = 1.

2. Three orthographic views and an isometric view of an object that has overall dimensions of 200 × 200 × 200 mm are to be shown in a drawing. Determine the optimal standard metric scale to be used for the ISO sheet sizes indicated below:

   a. A1.

   b. A3.

   c. A4.

3. The drawing shown in Figure 5.13 has many errors in the dimensioning. Sketch the orthographic views and show proper placement of dimensions.

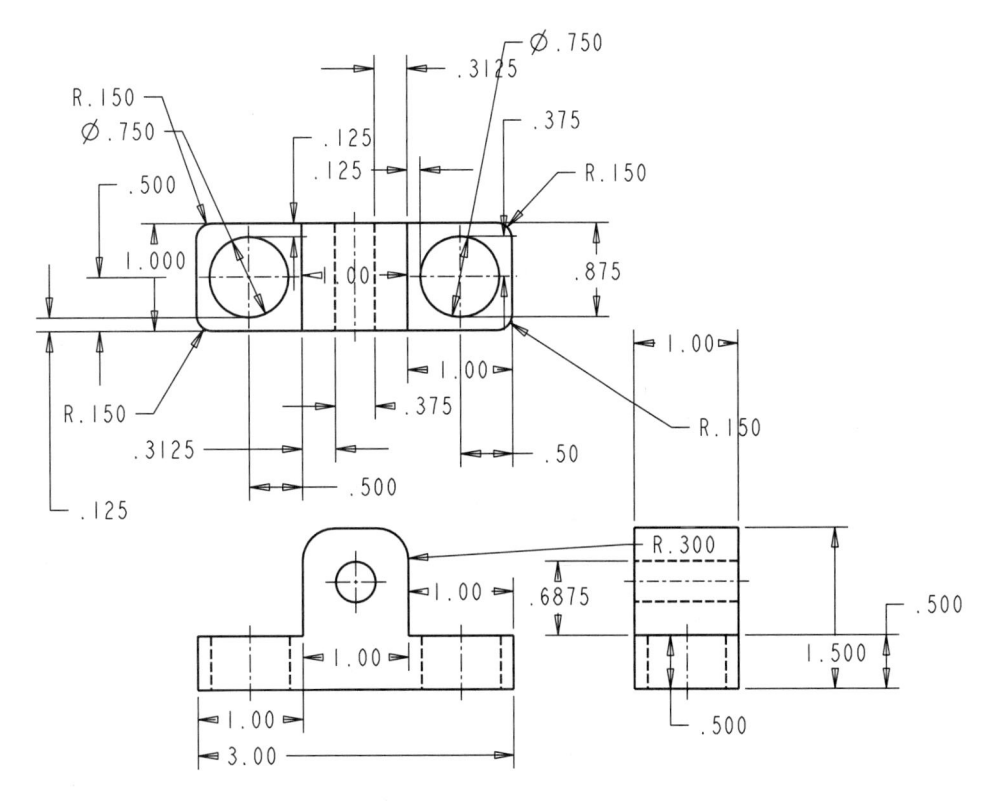

Figure 5.13.

4. The drawing shown in Figure 5.14 has many errors in the dimensioning. Sketch the orthographic views and show proper placement of dimensions. The M10 × 1.5 threaded hole requires a tap drill of 8.5 mm. Only two views are needed for this part.

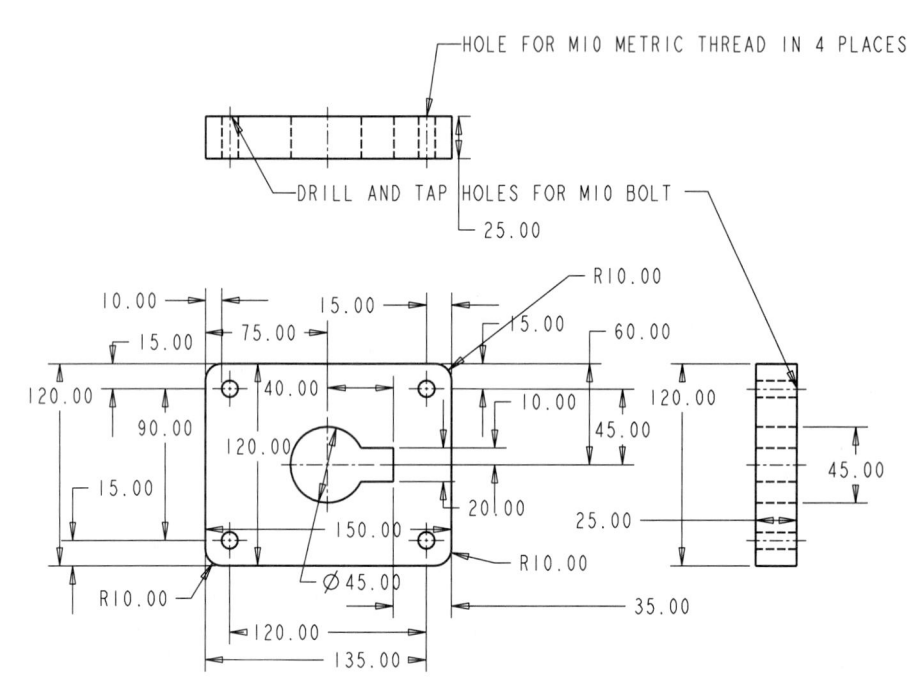

Figure 5.14.

5. Sketch the following section views for the object shown in Figure 5.15. The threaded holes extend from the top surface approximately halfway through the thickness of the part. The other holes extend through the part.

   a. A-A.
   b. B-B.

Isometric View   Top View

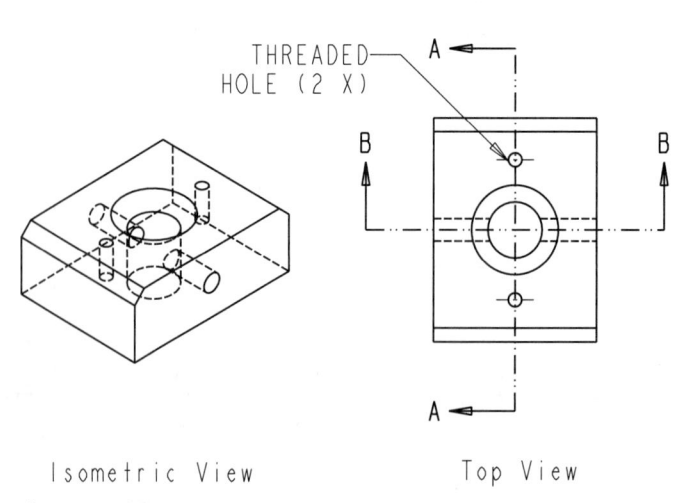

Figure 5.15.

6. Sketch the following section views for the object shown in Figure 5.16. All holes extend through the part.

   a. B-B.
   b. D-D.

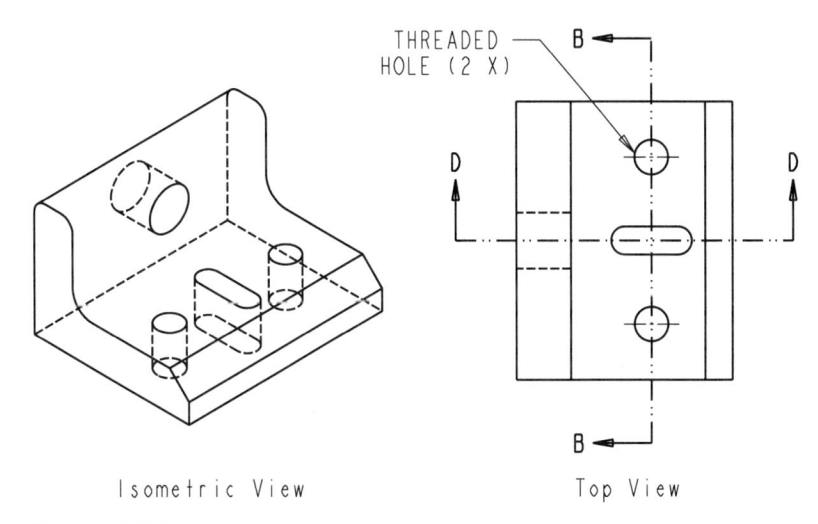

Figure 5.16.

7. Look up the following machine screw threads in an engineering graphics book (such as *Engineering Graphics* by Giesecke, et al), a machinists or engineering handbook (such as *Marks' Standard Handbook for Mechanical Engineers*), an industrial supply catalog (such as McMaster–Carr), or on the World Wide Web (search on "tap drill size"). Specify the series (UNC or UNF), tap drill size or number, number of threads per inch, pitch, and nominal diameter.

   a. 4-40.
   b. 6-32.
   c. 8-32.
   d. 3/8-24.
   e. 6-40.
   f. 1/4-28.

8. Look up the following machine screw threads in an engineering graphics book (such as *Engineering Graphics* by Giesecke, et al), a machinists or engineering handbook (such as *Marks' Standard Handbook for Mechanical Engineers*), an industrial supply catalog (such as McMaster–Carr), or on the World Wide Web (search on "tap drill size metric"). Specify the series (course or fine), tap drill size, number of threads per mm, and nominal size.

   a. M2 × 0.4.
   b. M12 × 1.75.
   c. M30 × 3.5.
   d. M30 × 2.
   e. M10 × 1.25.

6

Tolerances

OVERVIEW

Tolerances on dimensions are necessary to specify the acceptable variability in the dimension of a part. Tolerances can be indicated for all dimensions using a general note or they can be specified for each individual dimension. Tolerances are based on the function of the part or are used to assure that mating parts fit together. However, in many cases the manufacturing process determines the tolerance. The surface finish may also be specified on a drawing to indicate the roughness of the surface. To further minimize ambiguity in engineering drawings, geometric dimensioning and tolerancing is used to specify the form of a feature, such as flatness or roundness, or to indicate the ideal position of a feature.

SECTIONS

- 6.1 Why Tolerances?
- 6.2 Displaying Tolerances on Drawings
- 6.3 How to Determine Tolerances
- 6.4 Surface Finish
- 6.5 Geometric Dimensioning and Tolerancing

OBJECTIVES

After reading this chapter, you should be able to:

- Explain interchangeable parts
- Explain the necessity of tolerancing
- Read a general tolerance on a drawing
- Read limit dimensions and tolerances on individual dimensions
- Determine a tolerance stackup
- Determine the range of tolerances for manufacturing processes
- Read a surface finish specification
- Understand basic form tolerances
- Understand simple positional tolerances
- Read geometric dimensioning and tolerancing symbols

6.1 WHY TOLERANCES?

In 1815, it was hoped that muskets could be produced in a number of different U. S. Government and private armories so that the parts of the musket would be interchangeable. Before this time, muskets were individually handcrafted, so parts from one musket rarely fit another musket. Because engineering drawings were not commonly used, a number of "perfect" muskets were made according to master gages and jigs. *Gages* are devices used to check the individual dimensions of a part. For instance, the diameter of a shaft could be checked using a gage consisting of two holes: one of the maximum allowable diameter of the shaft and one of the minimum allowable diameter. If the shaft fit through the larger hole but not the smaller hole, then it was within the allowable dimensions for the shaft. *Jigs* are devices used to hold a workpiece and guide the tool to assure repeatable machining. The master gages and jigs were sent to the armories with the instruction that "no deviations from the pattern were to be allowed." In 1824, this concept of *interchangeable parts* was tested by disassembling a hundred rifles from several different armories, mixing the parts, and then reassembling at random. Most reassembled rifles worked as they should, proving the concept and value of interchangeable parts.

The interchangeability of parts is so common now that it is hard to imagine anything different. The concept allows parts made in various locations by different manufacturers to be successfully assembled and to function properly as an assembly. Although engineering drawings rather than "master parts" came into use in the mid-19th century, it was the two world wars in the 20th century that brought about the development of methods of showing the acceptable variation in a dimension on a drawing, or *tolerancing*

The need to control precisely the geometry of a part arises from the part's function. For instance, the cross-sectional geometry of an airfoil, such as the wing of a jet aircraft or a turbine blade, must be accurately controlled to assure aerodynamic efficiency. The more commonplace need for controlling geometry results from the requirement for parts to fit together. But it is quite difficult to make every part the exact size that is specified in a dimension because of slight differences in tool size, machine tool wear, human operator error, and other factors. As a consequence, tolerances are used in engineering drawings to specify the limit in the variation between mating parts and to provide guidelines to control the manufacturing process. The *tolerance* is the amount that a specific dimension is permitted to vary, or the difference between the maximum and minimum limits of a dimension.

Consider a dimension on a drawing specified as 3.750 ± .003. This dimension indicates that the part has a *nominal size*, or general size, of 3 3/4 inches (usually expressed as a common fraction). The *basic size*, or theoretical size for the application of the tolerance, is 3.750 inches (expressed as a decimal). The tolerance of the dimension is .006 inches. Thus, an acceptable machined part can have an *actual size* ranging from 3.747 to 3.753 inches, which are the minimum and maximum limits of the dimension. If the actual size is smaller than 3.747 or larger than 3.753, then the part is not acceptable.

Of course, it is desirable to make the part as close as possible to the basic size, sometimes called the *target size*. But the more accuracy needed in a machined part, the higher the manufacturing cost. Furthermore, the method of machining a part limits the tolerance that can be specified. For instance, the tolerance in drilling a .500 inch hole is between .002 and .005 depending on the quality of the drill press and drill bit. To require a tolerance smaller than .002 would require an additional machining process, such as reaming. Consequently, tolerances can play a large role in the cost of manufacturing a part. Therefore, the tolerances should not be specified tighter than necessary for the product to function properly.

An increased awareness of the importance of tolerances in manufacturing has led to an approach in which the deviation of the actual size of a part from the target size has a cost in terms of reworking of the part, scrap, customer dissatisfaction, and poor reliability. An approach to handle this is to design a product so that it is less sensitive to manufacturing tolerances. The idea is to design the part intelligently so that easy-to-manufacture tolerances, rather than exceptionally tight tolerances, are necessary. Consequently, proper fit between mating parts comes about because of good design, instead of depending on tight tolerances. This approach is known as *robust design*.

6.2 DISPLAYING TOLERANCES ON DRAWINGS

Every dimension on a drawing should have a tolerance. Tolerances can be displayed on engineering drawings in several ways depending on the situation.

General tolerances can be given in a note on the drawing or in the title block. An example would be the note ALL DIMENIONS TO BE HELD TO ± .003″. Thus, a dimension of .375 would have a minimum limit of .372 and a maximum limit of .378. Often general tolerances are specified in terms of the number of digits following the decimal point in a dimension. In this case, a note would appear on the drawing such as:

$$\begin{array}{c} \text{UNLESS OTHERWISE SPECIFIED} \\ \text{TOL:.XX} = \pm.010 \\ \text{.XXX} = \pm.005 \\ \text{HOLES} = \pm.002 \\ \text{ANGLES} = \pm.5 \text{ DEG} \end{array}$$

(6-1)

Thus, a dimension of .50 on a drawing would indicate an actual size between .490 and .510. A dimension of .500 would indicate an actual size between .495 and .505. For metric dimension, a similar form for general tolerances would be X.X METRIC = ±.08.

Two other methods of displaying tolerances on a drawing are shown in Figure 6.1. The maximum and minimum sizes are specified directly when using *limit dimensions* to specify the tolerance. Usually the upper limit is placed above the lower limit. While this method clearly shows the limits of the dimension, many engineers, designers, and machinists think in terms of the nominal or basic size plus a tolerance, neither of which are immediately evident using limit dimensions. Thus, a more common approach is to use *plus and minus dimensions* where the basic size is shown followed by the tolerance values. If the plus and minus tolerance values are identical, then the basic dimension is followed by a plus/minus sign and half of the numerical value of the tolerance. Otherwise, the plus and minus tolerances are indicated separately with the plus above the minus. A *bilateral tolerance*, such as the tolerance for the 1.500 inch dimension in Figure 6.1, varies in both directions from the basic size. A *unilateral tolerance*, such as the tolerance on the shaft in Figure 6.1, varies in only one direction from the basic size. In both cases, the number of digits to the right of the decimal point for the dimension should be the same as the number of digits to the right of the decimal point for the tolerance.

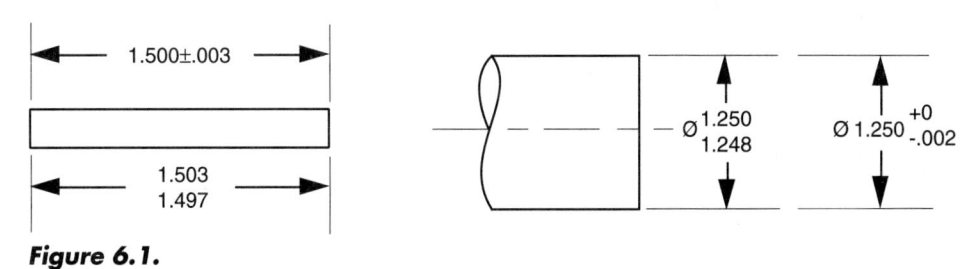

Figure 6.1.

Sometimes it is not necessary to specify both limits of a tolerance and only one limit dimension is needed. MIN or MAX is placed after a numerical dimension to indicate that it is a *single limit dimension*. The depth of holes, length of threads, and radii of corner rounds are often specified using single limit dimensions.

Many times several methods of representing tolerances are used on a single drawing. For instance, general tolerances may be listed in the title block and plus and minus dimensions or limit dimensions may be used for a few dimensions to which the general tolerances do not apply.

An important consideration in applying tolerances is the effect of one tolerance on another, especially because tolerances are cumulative. This is known as *tolerance accumulation* or *tolerance stackup* for a chain of dimensions. In the example shown in Figure 6.2, two dimensioning schemes are shown. Above the part, *chain dimensioning* shows the distance from one feature to the next. This would be an appropriate dimensioning scheme if the distance from one hole to the next were critical, because the distance between holes must be between .995 and 1.005. But if the distance of the holes from left edge of the part were critical, a large tolerance would accumulate. The third hole would be 2.985 from the left edge if all of the dimensions happened to be at the minimum limit of the dimension, or 3.015 from the left edge if all dimensions were at the maximum limit. Thus, the actual dimension of the third hole from the left edge using chain dimensioning is 3.000 ± .015. This tolerance accumulation could be avoided using *datum dimensioning*, where the dimensions are given with respect to a datum, in this case the left edge of the part as shown below the part in Figure 6.2. Using this scheme the third hole is specified to be between 2.995 and 3.005 from the left edge. Of course, the distance between the second and third holes could be as small as .990, if the second hole is at 2.005 and the third hole is at 2.995. The distance between the second and third holes could be as large as 1.010 if the holes are at 1.995 and 3.005. Consequently, the decision about which way to dimension the holes depends on whether the distance between holes is critical, or whether the distance from each hole to the left edge is critical.

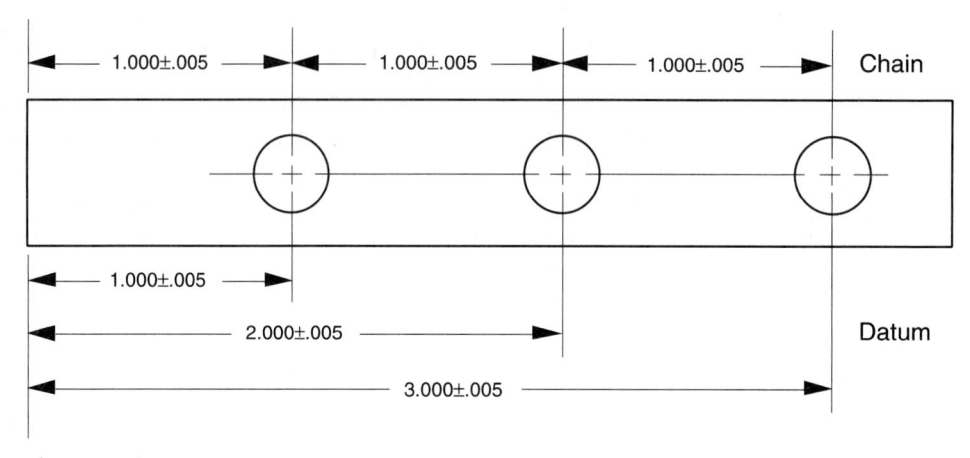

Figure 6.2.

6.3 HOW TO DETERMINE TOLERANCES

One of the most difficult aspects of dimensioning and tolerancing for inexperienced engineers and designers is to determine what tolerance to specify on a drawing. In some cases a specific tolerance is quite clear based on the function of the design. For instance, if a 3/4-inch shaft must rotate at moderate speed in a hole, the tolerances on the shaft and the hole can be determined based on standard classes of fits that are available in many engineering and machinist handbooks. In this particular case the hole could be between .7500 and .7512, while the shaft could have a diameter of .7484 to .7492. A shaft any bigger or a hole any smaller would result in too tight of a clearance for the shaft to rotate freely, or could result in the shaft being too big for the hole. A smaller shaft or larger hole than the specified limits would result in a sloppy fit between the shaft and the hole. The point here is not the details of how the tolerances are determined. That is too specific to individual cases for this discussion and is presented in many handbooks and texts. Instead, the point is that the function of the system that is being designed drives the tolerances. In this case, *the tolerances drive the manufacturing process* used to machine the hole and the shaft.

But more a more common situation is the opposite: *the manufacturing process drives the tolerance*. If the only machine tool that is available to form a hole is a drill press, the 3/4-inch hole that will result will be between .748 and .754 in diameter based on expected tolerances for the drilling operation. But this means that the hole in the example above would have too much variability to assure that the shaft rotates in it properly. Of course, if only one shaft-hole assembly was being made, the hole could be drilled first and measured. Then the shaft could be turned on a lathe to a diameter just slightly smaller to achieve the fit described above. But this approach will clearly not work for mass-produced, interchangeable parts. Either a more accurate machining method is needed to create the hole or the design must be altered to avoid drilling a tightly dimensioned hole altogether. (In this example, a standard-size, off-the-shelf bearing could be inserted into a larger drilled hole to mate with the shaft.)

When the manufacturing processes that are available drive the tolerances, the tolerances on the drawing must reflect the manufacturing process that is used to make the part, and the design must be altered to ensure that the part can be made with the available machining processes. Guidelines are available for the tolerance range of various common machining processes. These processes include the following:

> *Lapping*: A process to produce a very smooth and accurate surface by rubbing the surface of the workpiece against a mating form, often using a fine abrasive between the surfaces.
>
> *Honing*: A process in which a honing stone made of fine abrasive is used to form a surface on a workpiece.
>
> *Grinding*: A process to remove a small amount of material from a workpiece with a rapidly rotating grinding wheel.
>
> *Broaching*: A process similar to filing in which a cutting tool is reciprocated along its axis to remove a small amount of material from a workpiece.
>
> *Reaming*: Removal of a small amount of material from a drilled hole using a rotating cutting tool (reamer).
>
> *Turning*: Axisymmetric removal of material from a rotating workpiece by feeding a cutting tool into the workpiece using a lathe.
>
> *Boring*: Axisymmetric removal of material from a workpiece using a tool with a single cutting surface. Either the workpiece or the tool may be rotating.
>
> *Milling*: Removal of material by feeding workpiece into a rotating cutting tool.
>
> *Drilling*: Creating a hole by feeding the end of a sharpened cylindrical cutting tool (drill bit) into the material.
>
> *Stamping* and *Punching*: Cutting material using a punch and a die to shear the material, much like a paper hole-punch.

The tolerance typically depends on the size of the feature, as well as the quality of the machine tool, sharpness of the cutting tool, and skill of the operator. Figure 6.3 indicates the range of tolerances to be expected for various machining operations. To use this table, first consider the size of the feature in the left column. Then go below the sizes in the left column to the type of machining operation used. The bar associated with that process indicates the range of tolerances in the row associated with the feature size. For instance, consider a feature that has a dimension of 2 inches. If broaching were used to create that feature, the chart indicates that the total tolerance would be .0004 to .0015, depending on the machining conditions. Thus, this dimension could be specified as 2.0000 ± .0002 up to 2.0000 ± .0008 (rounding up). If the same piece was created on a milling machine, the tolerances would be .0025 to .010, indicating a dimension of 2.0000 ± .0013 up to 2.000 ± .005. If broaching were not available, we would have to be satisfied with the tolerances of a milling machine and adjust the design accordingly, if the milling tolerances are too large. If both broaching and milling were available, the criticality of the tolerance and the cost of manufacturing would drive the tolerance that would be specified on the drawing. (Broaching, a mass-production process, is usually more expensive than milling, so milling would be preferred if the milling tolerances are acceptable.) Tolerances for metric dimensioning can be determined by converting the values in Figure 6.3 to millimeters and rounding off the result to one less place to the right of the decimal point.

Size (in.)	Total Tolerance (in.)								
0.000-0.599	0.00015	0.0002	0.0003	0.0005	0.0008	0.0012	0.002	0.003	0.005
0.600-0.999	0.00015	0.00025	0.0004	0.0006	0.001	0.0015	0.0025	0.004	0.006
1.000-1.499	0.0002	0.0003	0.0005	0.0008	0.0012	0.002	0.003	0.005	0.008
1.500-2.799	0.00025	0.0004	0.0006	0.001	0.0015	0.0025	0.004	0.006	0.010
2.800-4.499	0.0003	0.0005	0.0008	0.0012	0.002	0.003	0.005	0.008	0.012
4.500-7.799	0.0004	0.0006	0.001	0.0015	0.0025	0.004	0.006	0.010	0.015
7.800-13.599	0.0005	0.0008	0.0012	0.002	0.003	0.005	0.008	0.012	0.025

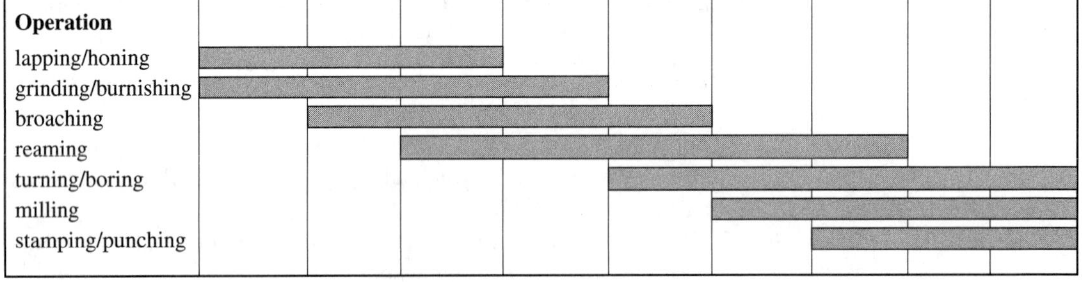

Operation

lapping/honing
grinding/burnishing
broaching
reaming
turning/boring
milling
stamping/punching

Figure 6.3.

Typical tolerances on drills depend on the drill size. Standard drill sizes are fractional in increments of 1/64-inch from 1/16-inch to 4-inch and according to an alphanumeric code. The coded drill sizes range from 80 (.0135 inch) to 1 (.2280 inch) and from A (.234 inch) to Z (.413 inch). The dimensions for the coded drill sizes are widely available in engineering and machinist handbooks. In addition, the proper drill sizes for drilling holes that will be tapped are listed in these handbooks. The standard tolerance for drilled holes is given in Table 6.1.

TABLE 6-1 Standard Tolerance For Drilled Holes

DRILL SIZE (IN.)	TOLERANCE (IN.)	
	Plus	Minus
0.0135 (80)–0.185 (13)	0.003	0.002
0.1875–0.246 (D)	0.004	0.002
0.250–0.750	0.005	0.002
0.7656–1.000	0.007	0.003
1.0156–2.000	0.010	0.004
2.0312–3.500	0.015	0.005

6.4 SURFACE FINISH

All surfaces are rough and irregular on a microscopic scale. The roughness can be described in terms of asperities (peaks or ridges) and valleys. The irregularities of a surface are classified into *roughness* and *waviness*. Roughness describes the small scale, somewhat random irregularity of a surface, while waviness describes a more regular variation in the surface that typically has a wavelength of 0.8 mm or more. The surface can be thought of as a roughness superimposed on waviness. The roughness and waviness can be measured in several ways, but the most common is a profilometer, which drags a fine-pointed stylus over the surface and records the height of the stylus. The roughness is measured in terms of the roughness average (R_a). The roughness average is defined as the average of the absolute values of the deviation of the local surface height from the average height of the surface. This is equivalent to half the average of the peak to valley height. In addition, the *lay*, or predominant direction of the marks left by the machine tools, may be important to describing the surface.

The surface finish resulting from several machining and forming processes are shown in Figure 6.4. This figure shows the range of possible surface roughness averages (R_a) in both micrometers (1μm = 10^{-6} m) and microinches (1μin = 10^{-6} in). For instance, surfaces formed by drilling are likely to have a roughness from 12.5 down to 0.8 μm (500 down to 32 μin). Typically, the roughness is in the middle of the specified range, about 6.3 down to 1.6 μm for the case of drilling. In addition to the machining processes described in Section 6.3, the following processes are also included in Figure 6.4:

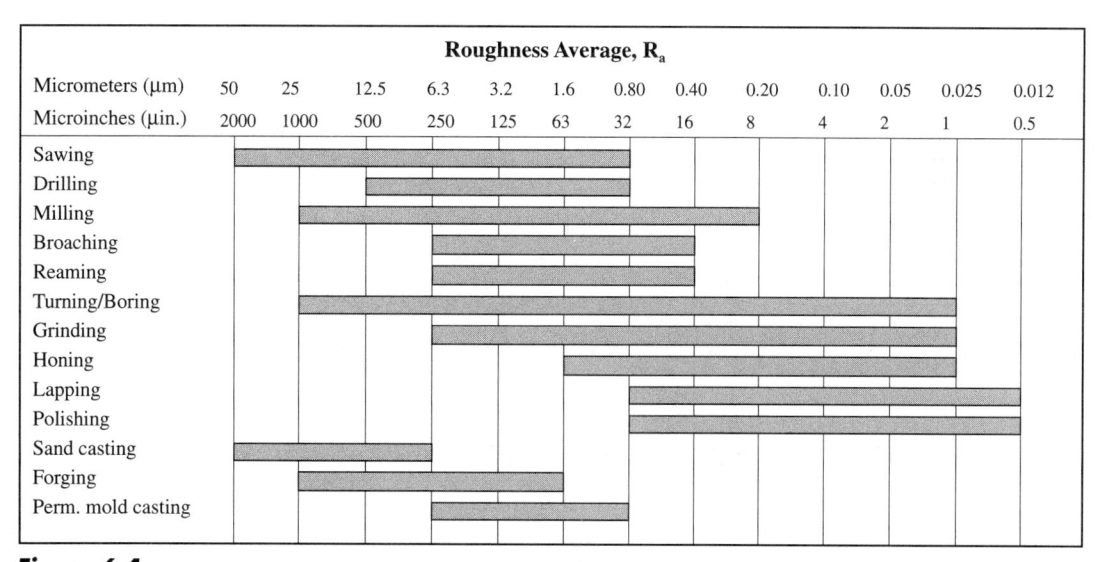

Figure 6.4.

Sawing: Cutting material by moving a toothed blade through the material.
Polishing: Removal of scratches and tool marks on a surface using a belt or rotating wheel of soft material such as cloth or leather, often with a fine abrasive.
Sand Casting: Forming a part by pouring molten metal into a mold made of sand.
Forging: Shaping of metal to a desired form by pressing or hammering, usually with the metal hot.
Permanent Mold Casting: Forming a part by pouring molten metal into a permanent mold.

To put the roughness averages in perspective, consider that a 0.10 µm roughness is a mirrorlike surface, free from visible marks of any kind. A 1.6 µm roughness is a high-quality, smooth machined finish. A 6.3 µm roughness is the ordinary finish from typical machining operations.

The surface finish may be specified on a drawing using the special symbols touching the surface or with a leader line to the surface as shown in Figure 6.5. For the top surface, the roughness is specified as 1.2 µm, corresponding to a high quality, smooth machine finish. The surface finish symbol on the right side of the object includes a horizontal bar that indicates that material removal by machining is required for this surface. The surface finish symbol on the left side of the object indicates that material removal from this side is prohibited. This surface must be that resulting from the original process used to produce the surface such as casting, forging, or injection molding.

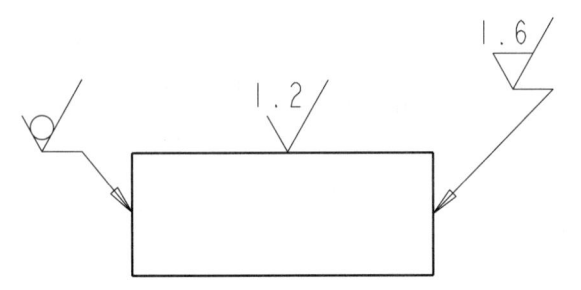

Figure 6.5.

Other notations can be added to the surface roughness symbol. A number to the left of the surface finish mark describes the material removal allowance indicating the amount of stock material to be removed by machining. A symbol to the right of the surface roughness mark indicates the lay, or predominant direction of machining marks: perpendicular, parallel, circular, or radial. Other notations can be used to specify the waviness. Details of surface finish notation are available in engineering and machinist handbooks.

6.5 GEOMETRIC DIMENSIONING AND TOLERANCING

Ideally, engineering drawings should be unambiguous in their interpretation. In many cases, stating dimensions and tolerances provides adequate information so that the part can be manufactured to ensure interchangeability of parts and optimal performance. But sometimes, the traditional dimensioning using the plus and minus scheme does not adequately describe the geometry of the part. For instance, consider the part shown in Figure 6.6. The dimension shown in the drawing could have several possible interpretations. For instance, the dimension at the top edge could be 14.9 mm while the dimension at the bottom edge could be 15.1 mm, resulting in a trapezoidal part instead of a rectangular part. Or the right edge of the part could be bowed so that the part is 15.1 mm wide at the top and bottom, but 14.9 mm wide at the middle. As a result, the part would not fit flush against another flat surface even though the part is 15 ± .1 mm wide at any particular location. In both of these cases, the geometry of the part is not described fully by the plus and minus tolerances on the dimension.

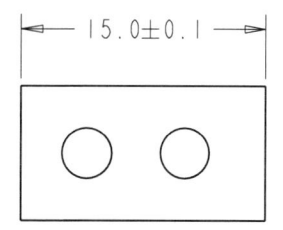

Figure 6.6.

A system known as *geometric dimensioning and tolerancing* is used to prevent such ambiguities in the design to present the design intent more clearly. Geometric dimensioning and tolerancing supplements the traditional dimensioning system with specifications that describe the geometry of the part to enhance the interpretation of the drawing. It is used to indicate dimensions critical to a part's function and ensure part interchangeability. Items related to geometric dimensioning and tolerancing are indicated on a drawing by a feature symbol consisting of a frame or box around information describing the geometry. A complete discussion of geometric dimensioning and tolerancing is beyond the scope of this book—there are entire books on the subject. Nevertheless, several of the key aspects and applications of geometric dimensioning and tolerancing are presented here to familiarize the reader with the basic terminology and symbols that are used.

The various types of tolerances and the characteristics that they describe are indicated in Figure 6.7. Each has its own symbol that is used on a drawing. Perhaps the easiest

	TYPE OF TOLERANCE	CHARACTERISTIC	SYMBOL
FOR INDIVIDUAL FEATURES	FORM	STRAIGHTNESS	—
		FLATNESS	▱
		CIRCULARITY (ROUNDNESS)	○
		CYLINDRICITY	⌭
FOR INDIVIDUAL OR RELATED FEATURES	PROFILE	PROFILE OF A LINE	⌒
		PROFILE OF A SURFACE	⌓
FOR RELATED FEATURES	ORIENTATION	ANGULARITY	∠
		PERPENDICULARITY	⊥
		PARALLELISM	∥
	LOCATION	POSITION	⌖
		CONCENTRICITY	◎
		SYMMETRY	═
	RUNOUT	CIRCULAR RUNOUT	↗
		TOTAL RUNOUT	↗↗

Figure 6.7.

type of geometric tolerances to understand are the *form tolerances*, which describe the shape of a single feature. For instance, straightness indicates the limits on how much a surface or axis can bow with respect to a straight line. Figure 6.8 shows the *straightness* specification for a surface. A feature control box with the straightness symbol (a horizontal bar)

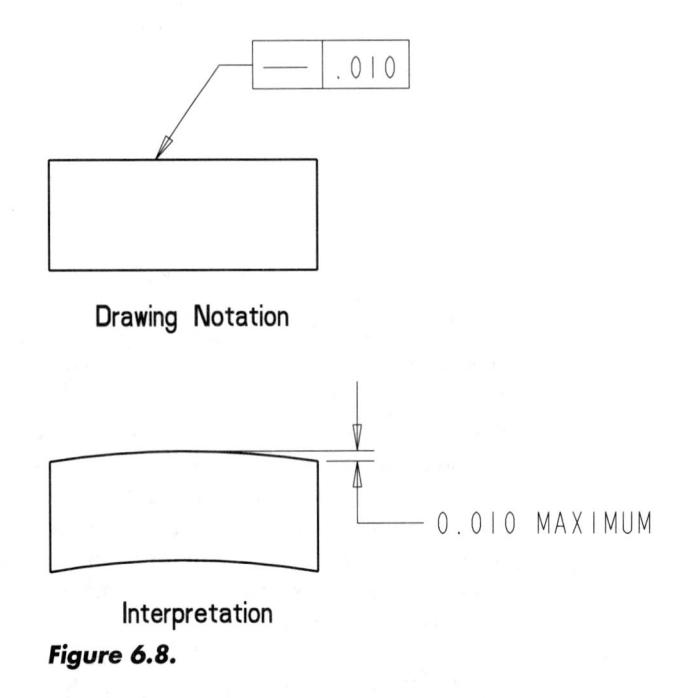

Figure 6.8.

and a numerical dimension are connected by a leader line to the surface. The interpretation is that the upper surface must be straight enough so that all points on the surface are within the tolerance zone of 0.010 mm. An example of a *flatness* specification is shown in Figure 6.9. The parallelogram symbol with a numerical dimension in a feature control box

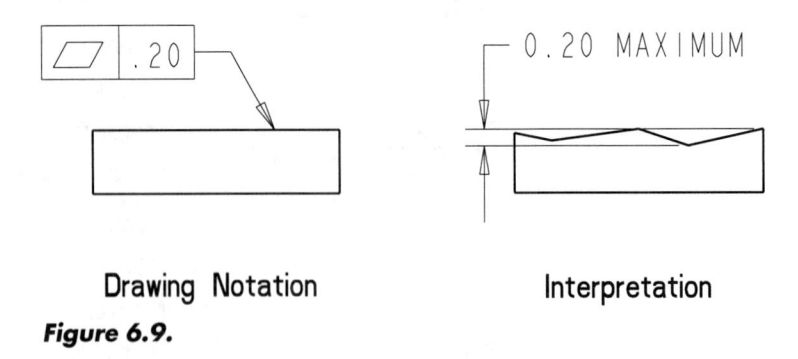

Figure 6.9.

indicates that the planar surface must lie between two parallel planes 0.20 mm apart. Other single feature specifications include *circularity* or *roundness* (for circular features to be within a tolerance zone defined by two concentric circles), and *cylindricity* (for cylindrical features to be within a tolerance zone defined by two concentric cylinders). Closely related to form tolerances are *profile tolerances*, which define the acceptable deviation of the outline of an object (profile) from that specified in the drawing. These tolerances define a zone on either side of the true profile shown on the drawing within which the line or surface should remain. Profile tolerances can be applied to either a line or a surface.

Orientation tolerances prescribe the relation between features. *Parallelism* controls the degree to which a surface is parallel to a reference plane, or datum plane. In Figure 6.10, the horizontal datum plane that coincides with the bottom edge of the part is identified with a box around B attached to the bottom edge. The parallelism fea-

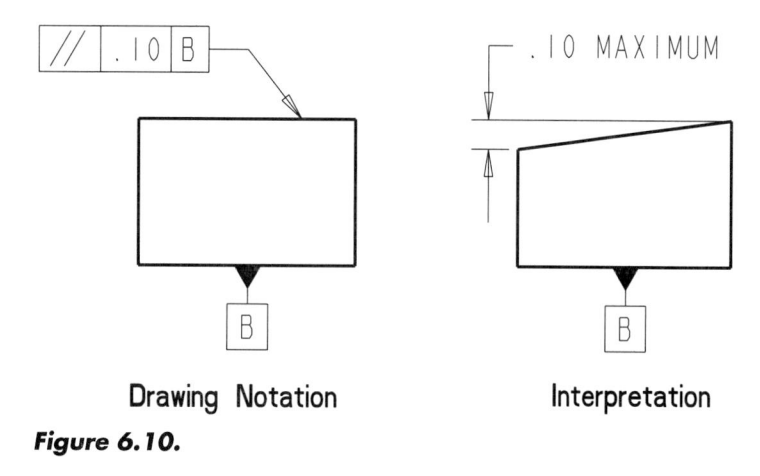

Figure 6.10.

ture symbol (parallel lines), the numerical tolerance, and the reference datum plane are included in a feature control box with a leader line to the surface to which the tolerance applies. In this case, the upper surface must be within .10 mm of parallel to datum plane B. *Perpendicularity* with respect to a datum plane is indicated in a similar way, as shown in Figure 6.11. Here the vertical plane must be perpendicular to datum plane A to

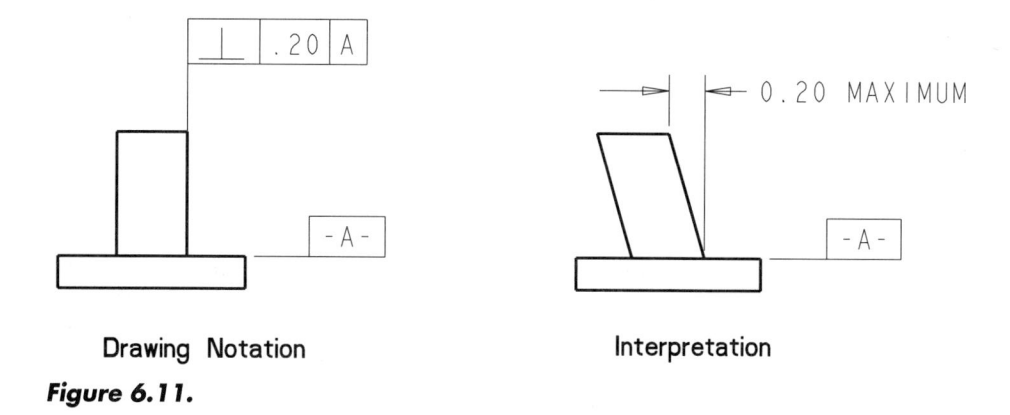

Figure 6.11.

within .20 mm. In this example, an alternative notation for specifying datum plane A is used. The datum plane symbol used in Figure 6.11 is common, but the datum plane symbol used in Figure 6.10 and Figure 6.12 is preferred. The datum plane symbol can be applied in several different ways as shown in Figure 6.10 and Figure 6.12. Several other relations between features can be specified, including the tolerance zone for *angularity* from a datum plane or axis, the tolerance zone for *concentricity* of a surface of revolution with respect to an axis, the tolerance zone for *symmetry* with respect to a datum plane, and the tolerance zone for *runout*, which describes the circularity with respect to an axis of revolution.

78 Chapter 6 Tolerances

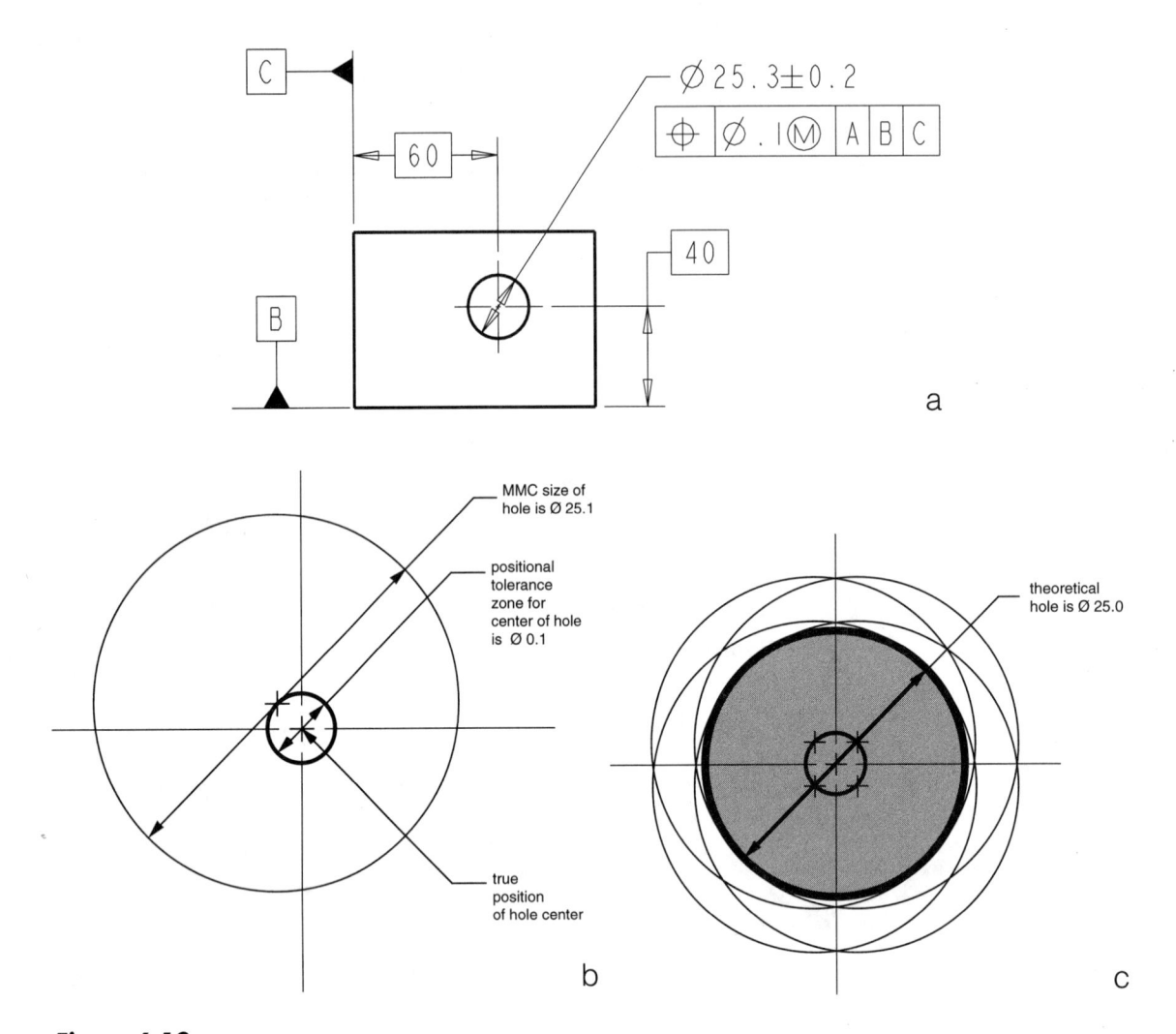

Figure 6.12.

The most difficult concept in geometric dimensioning and tolerancing is the concept of *positional tolerances*. It is demonstrated most easily in indicating the position of a hole, as shown in Figure 6.12a. In this case, the ultimate goal is to have a circular open area (hole) that is at least 25.0 mm in diameter with its center exactly at the position indicated in the drawing. Note that even though we want a 25.0 mm hole, the hole is dimensioned as 25.3 mm in the drawing. The reason for this will shortly become evident.

Rather than describing the position of the hole with a plus-or-minus tolerance, the ideal position of the hole is prescribed. The boxed dimensions indicate that they are basic *true-position dimensions*, which describe the perfect position of the center of the hole. Then the feature control box below the dimension of the hole shows how close the center of the hole should be to the true position. In this case, the hole should be within a circular tolerance zone with a diameter of .1 mm of the true position (indicated by the plus in the circle within the feature control box followed by ∅ .1). The reference datum planes are also included in the feature control box. Datum plane A is the surface of the part in the plane of the page. Datum planes B and C are references for the true-position dimensions of 40 and 60 mm. These three mutually perpendicular datum planes provide a reference point for the positional tolerance of the hole.

The positional tolerance zone for the center of the hole is shown in Figure 6.12b. The "true position of hole center" is based on the 40 and 60 mm dimensions. Around this center is a circle with a diameter of .1 mm (not drawn to scale in Figure 6.12b) that defines the "positional tolerance zone." The center of the actual hole must be within this circle. For instance, an acceptable hole would be a 25.1 mm diameter hole with its center on the edge of this circular tolerance zone, as shown in the Figure 6.12b.

The M in a circle in the feature control box in Figure 6.12a indicates that the .1 mm circular positional tolerance zone is applied at the *maximum material condition* (MMC). For a hole, the maximum material condition is the smallest acceptable dimension of the hole, because this is the situation in which the most material would be present. For the part in Figure 6.12a, the diameter of the hole for the maximum material condition is 25.1 mm, given a dimension of 25.3 ± .2 for the diameter of the hole. Thus, the circle representing a hole with its center on the edge of the positional tolerance zone has a diameter of 25.1 mm, as shown in Figure 6.12b. Of course, the 25.1 mm diameter hole could be anywhere within the positional tolerance zone or on its edge. Three more 25.1 mm diameter circles are drawn in Figure 6.12c in addition to the original 25.1 mm circle. The centers of all four circles are on the edge of the positional tolerance zone. Note that there is a circle that lies inside all of the 25.1 mm diameter circles, which we call the "theoretical hole." It is relatively straightforward to find the diameter of the theoretical hole, based on the positional tolerance zone for the center of the hole and the MMC size of the hole. The radius of the theoretical hole is the distance from the true position of the hole center to the nearest portion of any one of the MMC circles. Here the theoretical hole has a diameter of 25.0 mm. In fact the theoretical hole always has a diameter equal to the diameter for the maximum material condition minus the diameter of the positional tolerance (in this case, 25.1 mm − .1 mm = 25.0 mm).

The key result is that if the hole diameter is within the specified tolerance (25.3 ± .2 mm) and the center of the hole is within the positional tolerance zone (.1 mm), there will always be a circular opening in the part with a diameter of at least 25.0 mm. Now assume that the hole must mate with a circular pin that is positioned at the true center of the hole and is exactly 25.0 mm in diameter. Given the dimensioning scheme in Figure 6.12a, the pin will always align with the theoretical hole and fit through the theoretical hole, even if the actual hole is not positioned so its center is exactly at the true position of the hole center. Of course, the theoretical hole is based on the maximum material condition (smallest hole allowed). If the hole is at the large end of the acceptable tolerance, 25.5 mm in diameter, circles drawn in Figure 6.12c will be larger. But the theoretical hole will still be open for the pin. Likewise, if the circles drawn in Figure 6.12c are not on the edge of the positional tolerance zone, but inside of the zone, the entire theoretical hole will still be open for the pin. Thus, defining the position and diameter of the hole as in Figure 6.12a forces the existence of an open area (theoretical hole) that is at least 25.0 mm in diameter and centered at the true position of the hole center.

The maximum material condition reflects the practical aspect of assembly. Consider two parts that are bolted together with several bolts. If the bolt holes are not accurately positioned, not all of the bolts can be inserted. Either drilling out the holes to make them larger or using smaller diameter bolts fixes the problem. The idea of the maximum material condition is to capture these trade-offs. The maximum material condition indicates the inner boundary (the theoretical hole) for the actual hole. The boundary of the actual hole must always be outside of this inner boundary. In this way, a bolt or pin that is positioned at the true position of the hole will always fit through the actual hole. Likewise, the least material condition (LMC), or largest hole size, can be used to define a theoretical circle within which the actual hole will be. The boundary for the actual hole must lie inside of the theoretical circle defined by the LMC. Further information regarding geometric dimensioning and tolerancing is provided in handbooks and graphics textbooks.

KEY TERMS

Actual size
Basic size
Form tolerances
General tolerances
Geometric dimensioning and tolerancing
Limit dimensions
Maximum material condition
Nominal size
Orientation tolerances
Plus and minus dimensions
Positional tolerances
Roughness
Target size
Tolerance stackup
True-position dimensions

Problems

1. The part shown in Figure 6.13 is made using a very accurate milling machine by a skilled machinist. The hole is reamed. Specify plus and minus tolerances for each dimension assuming that it is critical to hold all dimensions as close as possible to those specified. Use four digits past the decimal point to specify the tolerances, even though typical practice would be to specify only three digits past the decimal point. Dimensions are in inches.

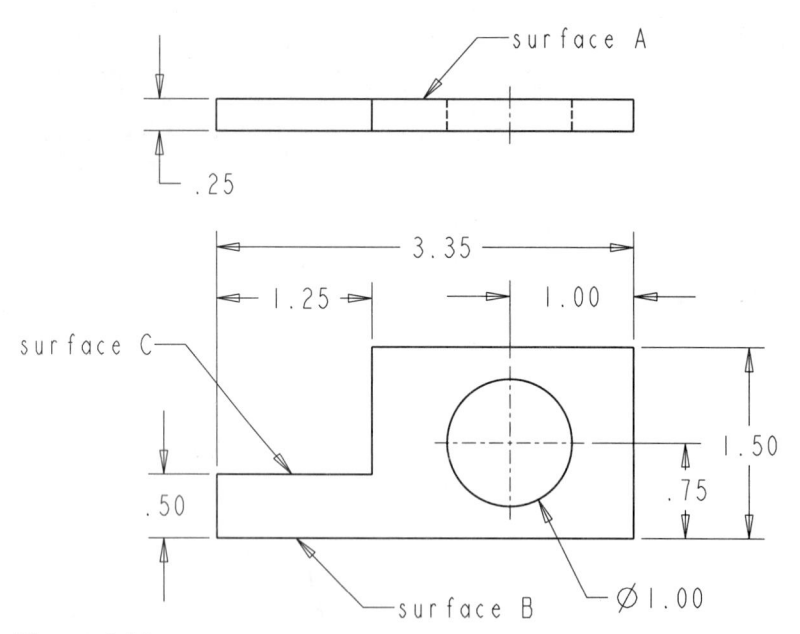

Figure 6.13.

2. The part shown in Figure 6.13 is made by an unskilled student using an old milling machine in a university student shop. The hole is drilled. Specify plus and minus tolerances for each dimension that are reasonable for these conditions. Use four digits past the decimal point to specify the tolerances, even though typical practice would be to specify only three digits past the decimal point. Assume dimensions are in inches.

3. Sketch a view of the object shown in Figure 6.14. Specify the horizontal dimension of the steps using chain dimensioning from the left end. The length of each horizontal section from left to right is .50, .75, 1.25, and .25 inches. Base the plus and minus dimensions on the worst expected tolerances for milling. What are the maximum and minimum dimensions for the overall horizontal length of the object? Use four digits past the decimal point to specify the tolerances, even though typical practice would be to specify only three digits past the decimal point.

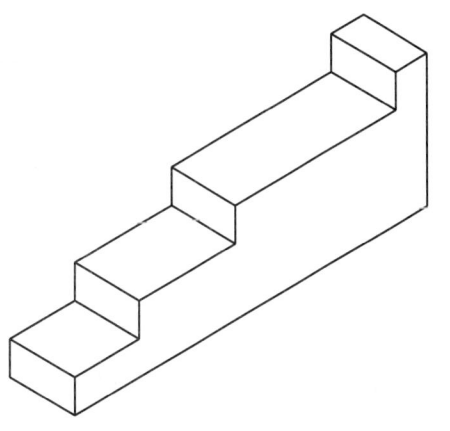

Figure 6.14.

4. Sketch a view of the object shown in Figure 6.14. Specify the horizontal dimension of the steps using datum dimensioning from the left end. The length of each horizontal section from left to right is .50, .75, 1.25, and .25 inches. Base the plus and minus dimensions on the best expected tolerances for milling. What are the maximum and minimum dimensions for the overall horizontal length of the object? Use four digits past the decimal point to specify the tolerances, even though typical practice would be to specify only three digits past the decimal point.

5. Sketch a side view of the shaft shown in Figure 6.15. Specify the horizontal dimension of each section of the shaft using chain dimensioning from the left end. The length of each horizontal section from left to right is .75, 2.00, .50, 1.25, and 3.00 inches. Base the plus and minus dimensions on the best expected tolerances for turning. What are the maximum and minimum dimensions for the overall horizontal length of the object? Use four digits past the decimal point to specify the tolerances, even though typical practice would be to specify only three digits past the decimal point.

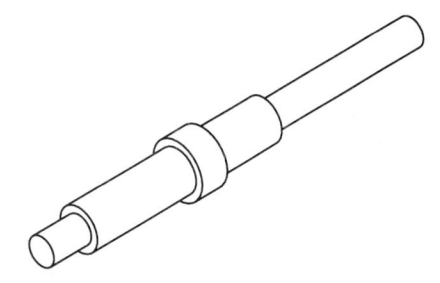

Figure 6.15.

6. Sketch a side view of the shaft shown in Figure 6.15. Specify the horizontal dimension of each section of the shaft using datum dimensioning from the left end. The length of each horizontal section from left to right is .75, 2.00, .50, 1.25, and 3.00 inches. Base the plus and minus dimensions on the worst expected tolerances for turning. What are the maximum and minimum dimensions for the length from the left end to the left edge of the largest diameter section? Use four digits past the decimal point to specify the tolerances, even though typical practice would be to specify only three digits past the decimal point.

7. Draw the appropriate surface finish symbol for:

   a. surface roughness typical for drilling.
   b. surface roughness typical for grinding.
   c. the best possible surface roughness for milling with material removal required.
   d. surface finish of 12.5 microns with material removal prohibited.

8. Sketch the appropriate Geometric Dimensioning and Tolerancing form tolerance for:

   a. an edge straight so that all points on the surface are within a tolerance zone of 0.020 mm.
   b. a surface that is flat to within 0.015 mm.
   c. a surface that is parallel to datum plane C to within 0.20 mm.
   d. a surface that is perpendicular to datum plane B to within 0.10 mm.
   e. a hole that is perpendicular to datum plane B to within 0.12 mm.

9. The part shown in Figure 6.16 is dimensioned in millimeters using the traditional coordinate method. Sketch the same views of the part and dimension using Geometric Dimensioning and Tolerancing procedures so the following functional requirements are met:

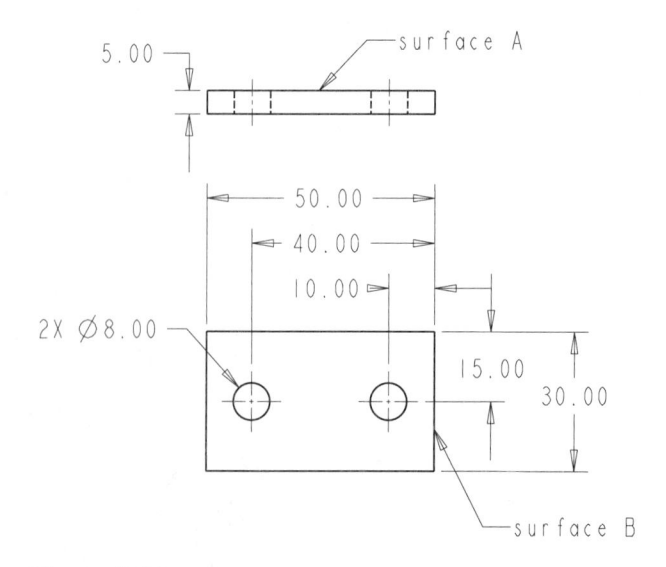

Figure 6.16.

- Set surfaces A and B as datum planes.
- Surface A is flat to within 0.15 mm.
- The upper edge in the front view from which the holes are located vertically is perpendicular to surface B to within 0.20 mm. Set this edge as datum plane C.
- The edge-forming surface A in the top view is straight to within 0.05 mm.
- The 10, 15, and 40 mm dimensions are true-position dimensions.
- The 5, 30, and 50 mm dimensions must be held to the tightest possible tolerance that can be achieved using milling. (Convert the dimensions to inches to determine the tolerances to two places after the decimal (round as necessary). Specify the tolerances as plus and minus dimensions in millimeters.)
- The holes have a tolerance of ±0.05mm. The theoretical hole size is 8.00mm.
- The holes are located within 0.15 mm of the true position at the Maximum Material Condition.

10. The part shown in Figure 6.13 is dimensioned in inches using the traditional coordinate method. Sketch the same views of the part and dimension using Geometric Dimensioning and Tolerancing procedures so the following functional requirements are met:

    - Set surfaces A and B as datum planes. Set the right edge as datum plane D.
    - Surface A is flat to within .005 inches.
    - The hole is perpendicular to surface A to within .010 inches. (Place this tolerance just below the dimension.)
    - Surface C is parallel to surface B to within .012 inches.
    - The .75 and 1.00 dimensions for the location of the hole are true-position dimensions.
    - The 1.50 and 3.35 dimensions must be held to the tightest possible tolerance that can be achieved using milling (use three digits past the decimal point for the tolerances).
    - The hole has a tolerance of ±.002. The theoretical hole size is 1.000.
    - The hole is located within .004 inches of the true position at the Maximum Material Condition.

11. The part shown in Figure 6.17 is dimensioned in millimeters using the traditional coordinate method. Sketch the same views of the part and dimension using Geometric Dimensioning and Tolerancing procedures so the following functional requirements are met:

    - Set surface A as a datum plane.
    - Surface B is flat to within 0.10 mm.
    - Surface B is parallel to surface A to within 0.05 mm.
    - The 40 and 100 dimensions need only be held to the loosest possible tolerance that can be achieved using milling. (Convert the dimensions to inches to determine the tolerances. Specify the tolerances as plus and minus dimensions in millimeters.)

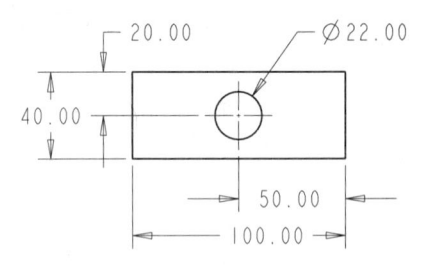

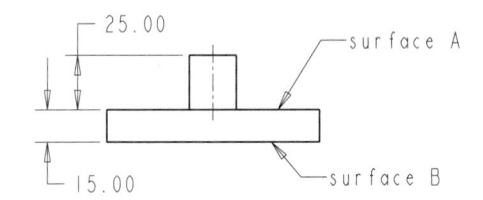

Figure 6.17.

References

ASME Y14.5M-1994, *Dimensioning and Tolerancing*. New York, NY: American Society of Mechanical Engineers, 1995.

G. R. Bertoline, *Introduction to Graphics Communications for Engineers*. New York, NY: McGraw-Hill, 1999.

P. J. Booker, *A History of Engineering Drawing*. London, England: Northgate Publishing Co. Ltd., 1979.

S. H. Chasen, Historical highlights of interactive computer graphics. *Mechanical Engineering*, November 1981, 32–41.

G. E. Dieter, *Engineering Design: A Materials and Processing Approach*. New York, NY: McGraw-Hill, 1991.

A. R. Eide, R. D. Jenison, L. H. Mashaw, L. L. Northup, and C. G. Sanders, *Engineering Graphics Fundamentals*. New York, NY: McGraw-Hill, 1985.

J. Encarnacao and E. G. Schlechtendahl, *Computer Aided Design*. Berlin, Germany: Springer-Verlag, 1983.

E. S. Ferguson, *Engineering and the Mind's Eye*. Cambridge MA: The MIT Press, 1992.

E. S. Ferguson, The mind's eye: Nonverbal thought in technology. *Science* 197:827–836, 1977.

F. E. Giesecke, A. Mitchell, H. C. Spencer, I. L. Hill, R. O. Loving, J. T. Dygdon, J. E. Novak, and S. Lockhart, *Engineering Graphics*. Upper Saddle River, NJ: Prentice Hall, 1998.

K. Hanks, *Rapid Viz: A New Method for the Rapid Visualization of Ideas*, Menlo Park, CA: Crisp Publications, 1990.

C. M. Hoffman, *Geometric and Solid Modeling*, San Mateo, CA: Morgan Kaufmann Publishers, Inc., 1989.

S. Pugh, *Total Design*, Reading, MA: Addison Wesley, 1990.

D. G. Ullman, *The Mechanical Design Process*, 2nd Edition. New York, NY: McGraw-Hill, 1997.

D. G. Ullman, S. Wood, and D. Craig, The importance of drawing in the mechanical design process. *Computers and graphics*, 14:263–274, 1990.

W. R. D. Wilson, Course Notes for ME B40: Introduction to Mechanical Design and Manufacturing. Northwestern University, March 1998.

7

Getting Started in SolidWorks®

OVERVIEW

In this chapter, you will begin modeling a pizza cutter. The guard, arm, and blade of the pizza cutter are modeled first, since they are the easiest parts to create. For all of these parts, you will sketch a two-dimensional section and then extrude it into the third dimension to create a base feature. Other features, such as rounds, cuts (holes), and chamfers, will be added to the base feature. You will also learn to spin the parts in order to view them from different angles.

7.1 INTRODUCTION AND REFERENCE

SolidWorks Corporation developed SolidWorks® as a three-dimensional, feature-based, solids modeling system for personal computers. Solids modeling represents objects on the computer as volumes, rather than just collections of edges and surfaces. Features are three-dimensional geometries with direct analogies to shapes that can be machined or manufactured, such as holes or rounds. Feature-based solids modeling creates and modifies the geometric shapes of an object in a way that represents common manufacturing processes. This makes SolidWorks a very powerful and effective tool for engineering design.

SECTIONS
- 7.1 Introduction and Reference
- 7.2 Modeling the Guard
- 7.3 Modeling the Arm
- 7.4 Modeling the Blade

OBJECTIVES

After working through this chapter, you should be able to

- Set the grid and units for a part
- Open, close, and save a part file
- Sketch lines and circles
- Sketch a two-dimensional section of a part
- Dimension a part
- Extrude a two-dimensional sketch
- Create chamfers, fillets, and rounds
- Create a circular cut (hole)
- Use the FeatureManager design tree to edit features
- Mirror features about a plane to create a symmetric part
- Undo or correct errors
- Use relations to define a part's geometry and
- View the part in various orientations

You are about to model an object that you have probably seen numerous times—a pizza cutter, shown in Figure 7.1. The pizza cutter consists of six different parts. You will model each of these parts, learning various commands along the way. Each concept that you learn will be applied again as you model subsequent parts and work on practice exercises. Once all the individual parts are modeled in Chapters 7 and 8, you will assemble those parts in Chapter 9 to create the pizza cutter model. In Chapter 10, you will create two-dimensional engineering drawings of one of the parts and the pizza cutter assembly.

Figure 7.1. Finished pizza cutter

As with other computer programs, SolidWorks organizes and stores data in files. Each file has a name followed by a period (dot) and an extension. The extension is a string of three letters that denote the type of file. For instance, a "part" file related to a handle could be named "handle.prt". The first part of the file name (handle) describes the contents of the file. The extension (.prt) indicates the type of file. There are several file types used in SolidWorks, but the most common file types and their extensions are

    Part files   .prt or .sldprt
    Assembly files   .asm or .sldasm
    Drawing files   .drw or .slddrw

Part files are the individual parts that are modeled. Part files contain all of the pertinent information about the part. Because SolidWorks is a solids modeling program, the virtual part on the screen will look very similar to the actual part once it has been manufactured. *Assembly files* are created from several individual part files that are assembled virtually (on the computer) to create the finished product. *Drawing files* are the two-dimensional engineering drawing representations of the part and assembly files. The drawings should contain all of the necessary information for the manufacture of the part, including dimensions, part tolerances, and so on.

The part file is the *driving* file for all other file types. The modeling procedure begins with part files. Subsequent assemblies and drawings are based on the original part files. One advantage of SolidWorks files is their dynamic links. Any change to a part file will automatically be updated to any corresponding assembly or drawing files. Therefore, drawing and assembly files must be able to find and access their corresponding part files in order for them to be opened. SolidWorks uses information embedded within the file and the filename to maintain these links automatically.

7.1.1 Starting SolidWorks

SolidWorks runs on computers running Microsoft Windows® operating systems. You open SolidWorks in the same way that you would start any other program:

Using the left mouse button, click "Start" (lower left corner of the screen), click "Programs", click **SolidWorks**, and then click on **SolidWorks** in the submenu;

or

Double-click on the **SolidWorks** icon on the screen with the left mouse button.

If the methods listed here do not work on your system, contact your system administrator for system-specific instructions.

7.1.2 Checking the Options Settings

The SolidWorks window that appears on the computer screen looks similar to the standard Microsoft Windows interface, as shown in Figure 7.2. The top line of the window is the Menu bar, from which menus of various operations can be opened. Below the Menu bar are the toolbars, which provide a variety of commonly used operations, or tools, with a single click of the mouse button. Toolbars can also extend down the right and left sides of the window. At this point, most items within the toolbars are grayed out. This indicates that they are not presently available for use. The main part of the window is the Graphics Window. This is where the model is displayed. Just below the Graphics Window is the Status bar, which displays information about the current operation. The Status bar should now indicate "Ready", since SolidWorks is ready to proceed.

This book will follow some basic conventions when presenting information and commands. All commands and tools will be in **bold**. Arrows (⇒) denote a move between menus or menu windows. Commands and tools can be in the Menu bar at the top of the window, in the toolbars at the top or sides of the window, or in the menus or windows that "pop up" on the screen. Commands and tools will be differentiated by the text style:

Plain	text indicates tools (accessible from either the toolbars or the Menu bar) or commands in the Menu bar.
Underlined	text indicates buttons in the toolbars.
Italic	text indicates commands or inputs in the windows or dialog boxes.

88 Chapter 7 Getting Started in SolidWorks ®

Figure 7.2. SolidWorks window

Before you begin, be sure that the SolidWorks settings match the ones used in this tutorial. This is done by setting up the appropriate options:

1. Click **Tools** in the Menu bar. Next, select **Options** (that is, **Tools ⇒ Options**).
   - The **Options** dialog box appears with the **System Options** tab visible, as shown in Figure 7.3.

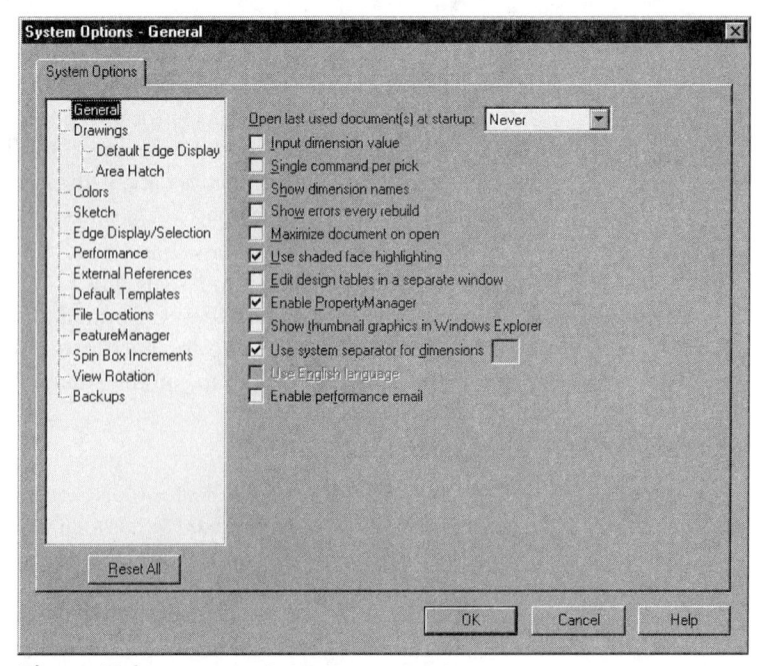

Figure 7.3. Options dialog box: General

2. Reset all of the preferences to the factory default by clicking **Reset All** followed by **Yes** in the confirmation dialog box. Click **OK** to close the dialog box.
3. To specify which toolbars are displayed on the screen, select **View** ⇒ **Toolbars** (i.e., select **Toolbars** from the **View** menu). Be sure that the **Standard**, **View**, **Features**, **Sketch**, **Sketch Relations**, and **Sketch Tools** toolbars are checked, as shown in Figure 7.4. If they are not, click on each of these items until all are checked. If other toolbars are checked, click on them to uncheck them. It may be necessary to select **View**, then **Toolbars** again to display the menu after checking (or unchecking) an item.

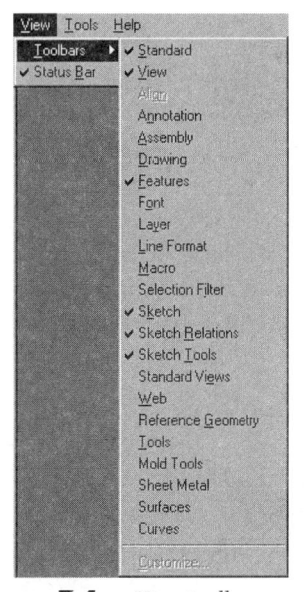

Figure 7.4. View toolbars menu

The selected toolbars are frequently used in SolidWorks. Each is described next:

- The **Standard** toolbar contains the usual commands available for manipulating files (Open, Save, Print, and so on), editing documents (Cut, Copy, Paste), and accessing Help.
- The **View** toolbar contains tools to orient and rescale the view of a part.
- The **Features** toolbar contains tools that modify sketches and existing features of a part.
- The **Sketch** toolbar contains tools to set up and manipulate a sketch of a cross section.
- The **Sketch Relations** toolbar contains tools for constraining elements of a sketch by using dimensions or relations.
- The **Sketch Tools** toolbar contains tools to draw lines, circles, rectangles, arcs, and so on.

You can find these toolbars around the Graphics Window by checking and unchecking them in the **View** ⇒ **Toolbars** menu. The toolbars will appear as you check them and disappear as you uncheck them. Currently, most of the items in the toolbars are grayed out, since they are unusable. They will become active when they are available for use.

7.1.3 Getting Help

If you have questions while you are using SolidWorks, you can find answers in several ways:

- Click **SolidWorks Help Topics** in the **Help** menu bar.
- Move the cursor over a toolbar button to see the name of the tool.
- Move the cursor over buttons or click menu items. The Status bar at the bottom of the SolidWorks window will provide a brief description of the function.
- Click the ***Help*** button in a dialog box.
- Refer to the *SolidWorks User's Guide*, by SolidWorks Corporation, for detailed information.

7.2 MODELING THE GUARD

The first part that you will model is the guard of the pizza cutter, shown in Figure 7.5. The guard is between the handle and the cutting blade of the pizza cutter. It is designed to prevent the user's hand from slipping from the handle onto the sharp blade while slicing a pizza. The guard would be fabricated from stainless steel sheet metal. Do not begin working on this part until you have set the ***Options*** as described in the previous section.

Figure 7.5.   Finished guard

7.2.1 Creating a New Part

1. With the SolidWorks window open, select **File ⇒ New** in the Menu bar, or click the <u>**New**</u> button (a blank-sheet icon) in the <u>**Standard**</u> toolbar.
2. The ***New SolidWorks Document*** dialog box appears. You will be modeling a new part. If *Part* is already highlighted, click **OK**. If it is not highlighted, click *Part*, then **OK**. A new window appears with the name **Part1**, as shown in Figure 7.6. On the left side of the window is the FeatureManager design tree. It contains a list of the features that have been created so far. Every

new part starts with six features: annotations, lighting, three datum planes, and an origin. The datum planes are three mutually perpendicular planes that are created in space as references for constructing features of the part that you are modeling. You might think of them as the x–y, y–z, and x–z planes in a three-dimensional Cartesian coordinate system. The planes are used to locate the features of the part within the virtual environment of SolidWorks. The three planes intersect at the origin, which is the intersection of the tails of the two arrows in the center of the Graphics Window. As the part is modeled, the features that are created will appear in the FeatureManager design tree. These features can be highlighted or modified by clicking on them in the FeatureManager design tree. For example, click on a plane or the origin in the FeatureManager design tree to highlight these items. **Plane1** is the plane of the screen, **Plane2** is the horizontal plane perpendicular to the screen, and **Plane3** is the vertical plane perpendicular to the screen. Finish by clicking on **Part1** in the FeatureManager design tree, so that no plane is highlighted.

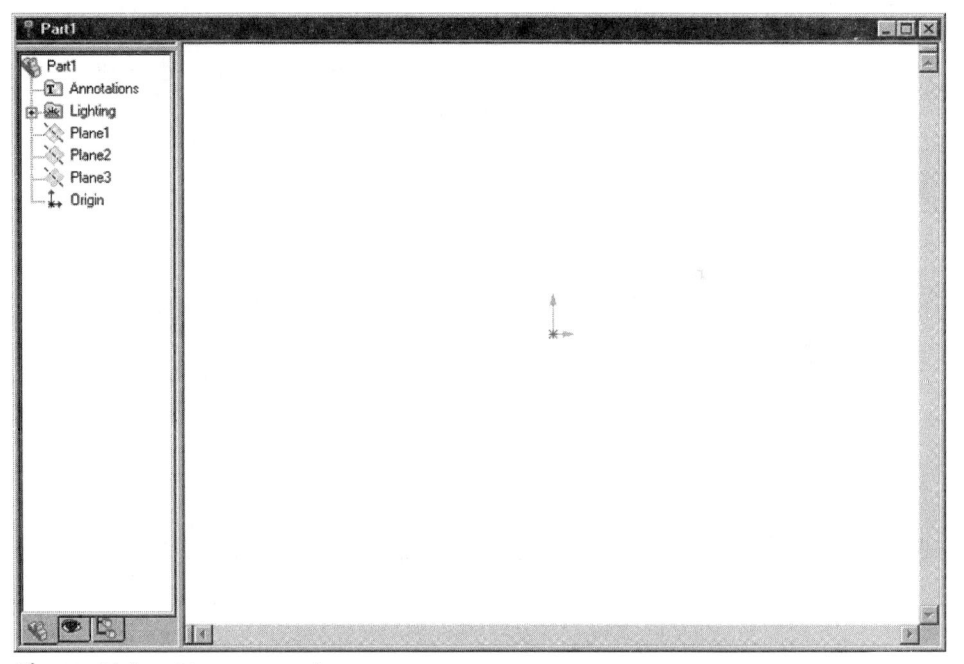

Figure 7.6. New part window

7.2.2 Sketching

Every part begins as a cross section sketched on a two-dimensional plane. Once a sketch is made, it is extruded or revolved into the third dimension to create a three-dimensional object. This is the base feature of the part. For the guard, a series of lines representing the shape of the guard's edge will be sketched. Then, a depth will be specified to extrude the guard perpendicular to the sketching plane.

1. The **Sketch** toolbar, shown in Figure 7.7, has tools to set up and manipulate a sketch of a cross section. Find the **Sketch** toolbar. Move the cursor over each

of the tools, but do not click on any of the tools. The ToolTips should appear, displaying the name of each tool. The sketch tools are described below.

- **Select** highlights sketch entities, drags sketch entities and endpoints, and modifies dimension values.
- **Grid** activates the ***Grid/Snap*** field of the ***Document Properties*** dialog box to change the sketching environment.
- **Sketch** opens and closes sketches as a part is created.
- **3D Sketch** activates three-dimensional sketching.
- **Dimension** adds dimensions to sketch entities.

Figure 7.7.   Sketch toolbar

2. To set the units and grid size that will be used, click the **Grid** toolbar button with the left mouse button.

   - Click ***Units*** on the left side of the dialog box to set the units. This tutorial will use inches, so set the ***Length unit*** to inches. To do this, click on the inverted triangle to the right of the box that says "Inches", as shown in Figure 7.8, so that the possible units are listed. Click on ***Inches*** so that it appears in the box. This tutorial will work with three decimal places. If necessary, change the ***Decimal places*** to "3" by clicking just after the number in the ***Decimal places*** box. Use the backspace key to delete the number that is shown, and then type in "3". The dimensions will be displayed as fractions, so click the button next to ***Fractions***. Change the ***Denominator*** to "32" so that fractions as small as 1/32 will be displayed. Set the ***Angular Unit*** to ***Degrees*** with "0" ***Decimal places***.
   - Click ***Grid/Snap***, on the left side of the dialog box to control the grid that will appear on the screen when a cross-section is sketched. Be sure that all three boxes under ***Grid*** are checked, as shown in Figure 7.8. The spacing will be 1 inch between major lines of the grid. Adjust the grid spacing to *1* by entering the value to the right of the ***Major grid spacing*** box.
   - ***Snap*** controls how the lines that are sketched are related to the grid. The points that are sketched should "snap" to the nearest intersection of grid lines when they are close. Be sure the ***Snap to points*** box is checked.
   - Be sure that all of the settings in the ***Document Properties*** tab match those shown in Figure 7.8. Then, click **OK** at the bottom of the dialog box to accept the values.

Section 7.2 Modeling the Guard   93

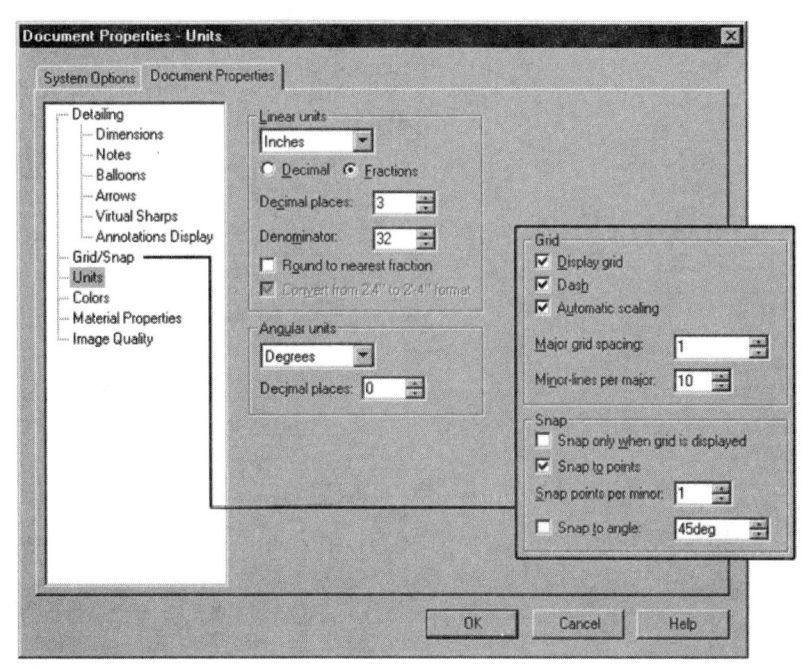

Figure 7.8.   Options dialog box: units

3. Open a new sketch by selecting **Insert ⇒ Sketch**, or by clicking the **Sketch** button (a pencil drawing a line) in the **Sketch** toolbar. (Note that, for most commands in SolidWorks, it is possible to implement the command from either the Menu bar or the toolbars.) A grid should appear on the screen, as shown in Figure 7.9, indicating that the sketch mode is active. The window's name changes to **Sketch1 of Part1**. In the bottom right of the screen, the Status bar reads **Editing Sketch**. You are now ready to sketch in **Plane1**.

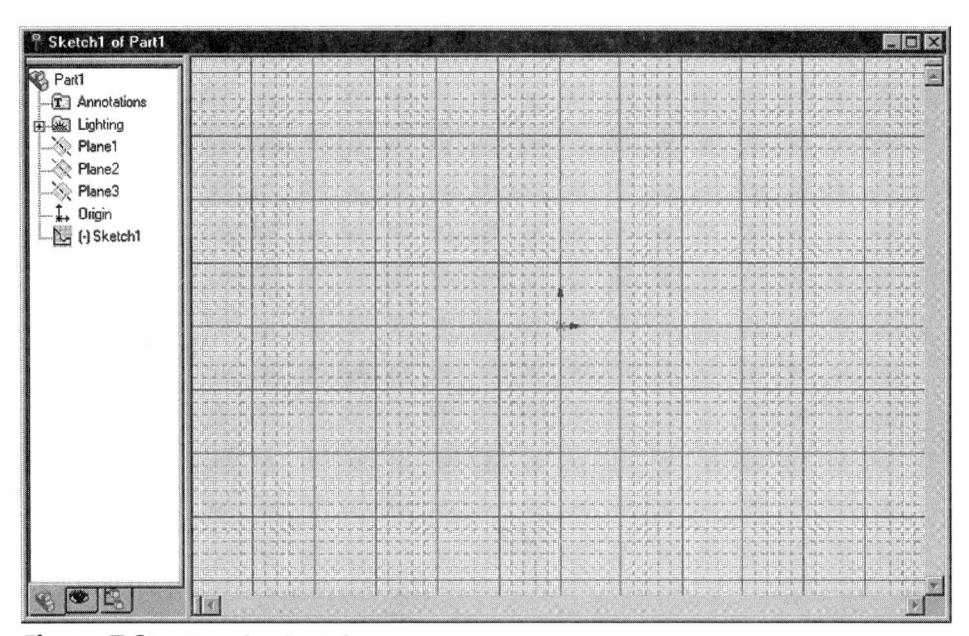

Figure 7.9.   New sketch window

4. The **Sketch Tools** toolbar, shown in Figure 7.10, contains tools to create and modify two-dimensional features called Sketch Entities. Sketch Entities are items that can be drawn on the sketch. The following Sketch Entities and Sketch Tools are available:

Figure 7.10.   Sketch Tools

- **Point** creates a reference point that is used for constructing other sketch entities.
- **Line** creates a straight line.
- **Rectangle** creates a rectangle.
- **Polygon** creates a polygon.
- **Centerline** creates a reference line that is used for constructing other sketch entities.
- **Circle** creates a circle.
- **Tangent Arc** creates a circular arc tangent to an existing sketch entity.
- **Centerpoint Arc** creates a circular arc from a centerpoint, a start point, and an end point.
- **3 Pt Arc** creates a circular arc through three points.
- **Spline** creates a curved line that is not a circular arc.
- **Trim** removes a portion of a line or curve.
- **Extend** makes a sketch segment longer.
- **Fillet** creates a tangent arc between two sketch entities by rounding an inside corner or outside corner.
- **Convert Entities** creates a sketch entity by projecting an edge, curve, or contour onto the sketch plane.
- **Offset Entities** creates a sketch curve that is offset from a selected sketch entity by a specified distance.
- **Intersection curve** opens a sketch and creates a sketched curve at the intersection of a plane and a surface.

- **Linear Sketch Step and Repeat** creates a linear pattern of sketch entities.
- **Circular Sketch Step and Repeat** creates a circular pattern of sketch entities.

Move the cursor over each of the tools to display its function, but do not click on the tool. Note the description of each tool in the Status bar at the bottom of the SolidWorks window. Some of these tools may not be included in the toolbar, depending on how it was previously set up. All tools are available in the **Tools** Menu.

5. Begin by sketching a line. Select the **Line** tool (an angled line) in the **Sketch Tools** toolbar by clicking on it, or select **Tools** ⇒ **Sketch Entity** ⇒ **Line** in the Menu bar. Bring the cursor over the grid. Notice that the cursor has changed to a pencil with a straight line next to it, indicating that the **Line** tool is active.

6. When you bring the cursor directly over the origin (at the tails of the two red arrows in the center of the grid), a square replaces the line near the cursor. This means that SolidWorks will snap one end of the line to the origin if you click there. With the square visible on the cursor, click *and hold* the left mouse button. Then, drag a line to the right, as shown in Figure 7.11. As you move away from the origin, the "H" below the pencil signifies that the line is horizontal. When the line looks similar to the one in Figure 7.11, release the mouse button. The line does not need to be exactly the same length as the line in the figure. If the line does not look like the one in Figure 7.11 click the **Select** button (an arrow) in the **Sketch** toolbar. Then, select the line by clicking on it with the left mouse button. (It will turn green.) On the keyboard, hit the delete key to delete the line. Using the **Line** tool, redraw the horizontal line from the origin. At this point, you need not dimension the line or draw it exactly the length shown in the figure. All of the sketch entities will be created, and later the appropriate dimensions will be added. This is known as constraint-based modeling.

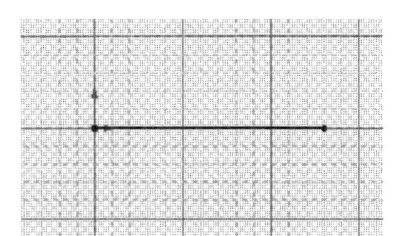

Figure 7.11. Horizontal line drawn

7. The **Line** tool should still be active. If it is not, click the **Line** button in the **Sketch Entities** toolbar. Draw a second line that starts from the right end of the first line and is angled upward and to the right. To do this, bring the cursor over the right endpoint of the first line so that the square on the cursor appears. Click and hold the left mouse button and drag the line up and to the right, as shown in Figure 7.12. This time the "H" will not appear, since this is not a horizontal line. Let go of the mouse button to finish the line.

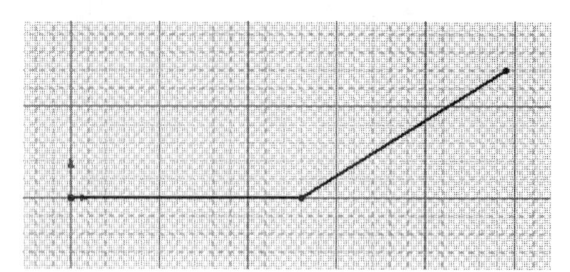

Figure 7.12. Angled line drawn

8. Complete the sketch shown in Figure 7.13 by repeating steps 6 and 7 to draw two lines on the left side of the origin. The first line drawn should be horizontal, and the second should go down and to the left. The sketched lines represent the contour of the guard's edge. These lines will be extruded into the sketching plane to create the three-dimensional base feature of the guard.

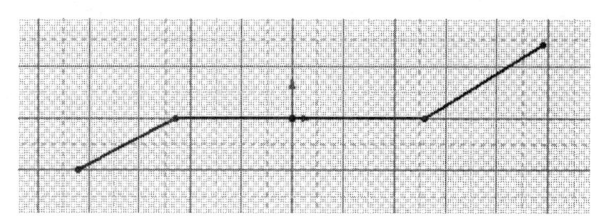

Figure 7.13. Other lines drawn

7.2.3 Dimensioning the Sketch

1. As noted on the right side of the Status bar at the bottom of the window, the sketch is *Under Defined*, because the dimensions have not been specified. The sketch will be *Fully Defined* when dimensions are added to "constrain" the geometry. The simplest dimension is the length of a line. Dimensions in SolidWorks can extend from any sketch entity to another. For example, the length of a line can be defined by dimensioning from one endpoint of the line to the other. Dimensions can also be assigned to a particular sketch entity. For example, a line can be dimensioned by specifying its length, and a circle can be dimensioned by specifying its diameter. Click on the **Dimension** button (a dimensioned line) in the **Sketch** toolbar or **Tools** ⇒ **Dimensions** ⇒ **Parallel**.

2. With the **Dimension** tool active, bring the cursor over the right endpoint of the first horizontal line that was drawn. Notice that a circle appears next to the cursor. Click the point. The point turns green.

3. Click on the Origin point. It also turns green, and a blue dimension box appears. The dimension box moves with the cursor. Click on the grid above the line with the left mouse button to place it in the position shown in Figure 7.14. Right now, the value of the dimension is not critical. The number in the dimension box represents the current length of the line. The number is probably not the same as the dimension in the figure, but it will be changed shortly.

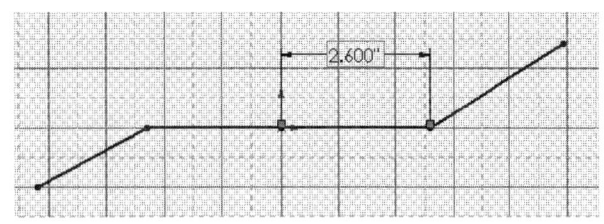

Figure 7.14. Horizontal dimension placed

4. The **Dimension** tool should still be active. Now, dimension the other horizontal line in the same manner as the first. Place the dimension at a convenient location above the line. Figure 7.17 shows this dimension, and the other dimensions that will be created shortly.

5. With the **Dimension** tool active, click on the two endpoints of the angled line on the right side. It does not matter which endpoint is selected first. Before you place the dimension box, move the cursor around. Depending on where the cursor is positioned, three different choices for the dimension appear. These different types of dimensions are described below and shown in Figure 7.15.

   - A *vertical* dimension, which defines the vertical distance between two points, appears when the cursor is to the right or left of both endpoints *and* vertically between them.

   - A *horizontal* dimension, which defines the horizontal distance between two points, appears when the cursor is above or below both endpoints *and* horizontally between them.

   - An *aligned* dimension, which represents the actual length of the line, appears when the cursor is close to the line, and at several other positions.

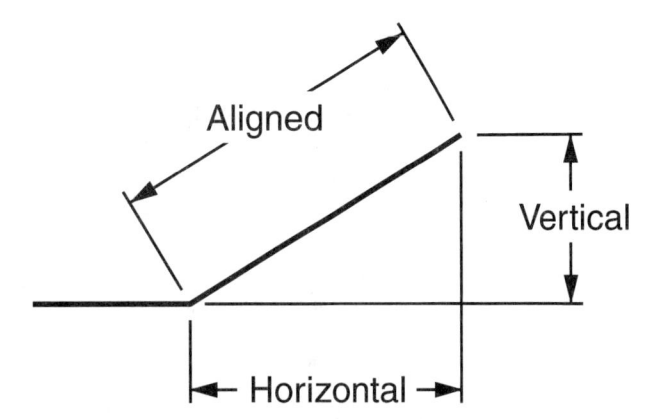

Figure 7.15. Different ways to dimension an angled line

Create a *vertical* dimension by placing the dimension box both between and to the right of both points. This dimension should be similar to the one shown in Figure 7.17, although the numerical value may not match exactly. If you placed a dimension that you do not want, use the **Select** tool (**Select** button in **Sketch** toolbar) to highlight the dimension. Then, delete it using the delete key on the keyboard. Redo the dimension with the **Dimension** tool.

6. Make another vertical dimension for the left angled line.

7. Dimensions can also be used to define angular values. Be sure that the **Dimension** tool is still active. Click on the angled line on the right, and then on the horizontal line it touches. When placing an angular dimension, there are three options to choose from, as shown in Figure 7.16. Place the dimension representing the angle of the line above the horizontal (Φ in Figure 7.16). To do this, click the mouse button when the dimension appears where Φ is shown in Figure 7.16. An angular dimension will appear, with degrees as the units.

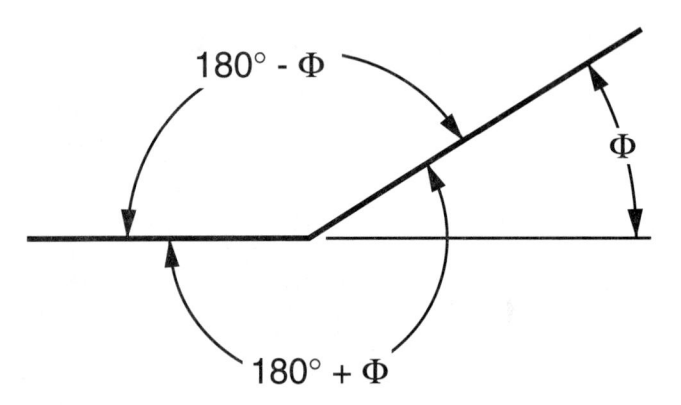

Figure 7.16. Different ways to dimension an angle

8. Once you have placed the dimension, activate the **Select** tool. Click and drag the angular dimension value to move it to a more convenient location, as shown in Figure 7.17.

9. At this point, the bottom right corner of the Status bar displays *Under Defined*, because more dimensions are needed in order to unambiguously represent the shape in the sketch. SolidWorks helps the user to know what dimensions are still needed by highlighting undefined sketch entities in blue. In this case, the left angled line is blue, because the angle must still be dimensioned with respect to the horizontal. Use the **Dimension** tool to dimension the angle of the blue angled line. After adding this dimension, all of the lines should be black. This color change indicates that the sketch is fully defined. That is, the dimensions and constraints completely describe the geometry of the sketch. The bottom right corner of the Status bar displays *Fully Defined*. The screen should look similar to Figure 7.17. The values of the dimensions may be different.

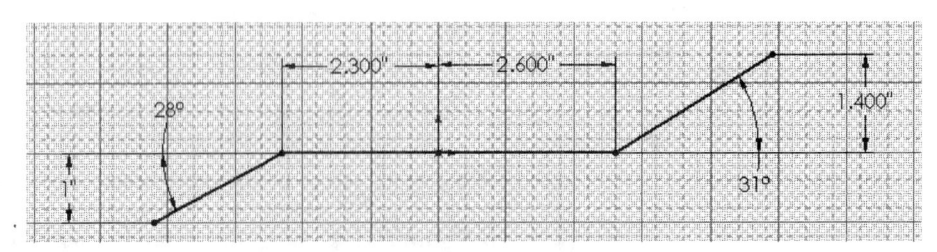

Figure 7.17. Dimensions placed

7.2.4 Changing Values of the Dimensions

In SolidWorks, changing the value of a dimension changes the properties of the sketch entity to which the dimension refers. Dimensions, along with constraints (parallel, horizontal), define the sketch. At this point, you must change the values of the dimensions to those that describe the shape of the edge of the guard.

1. Activate the **Select** tool by clicking the **Select** button in the **Sketch** toolbar. Double click the left mouse button on one of the linear dimensions that define the length of the horizontal lines. In the **Modify** dialog box, shown in Figure 7.18, type in the value **1.25**. Then, click the green check button in the **Modify** dialog box, or hit Enter on the keyboard. The length of the line is changed and the value shows "1 1/4". The units need not be specified while changing a dimension. The units were already set in the **Document Properties** dialog box. In SolidWorks, you can enter equations that represent dimension values. For example, the above dimension could be entered as "1+1/4" or "1 1/4" instead of 1.25.

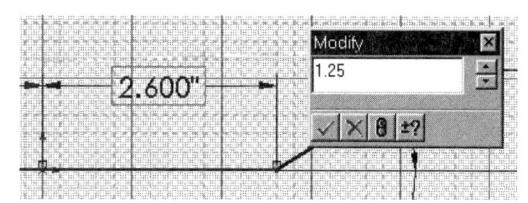

Figure 7.18.   Modify dialog box

2. Angular dimensions can be changed in the same way as linear dimensions. With the **Select** tool active, double click on one of the angular dimensions. Set the value to **30**. After you hit Enter, the angle automatically updates to 30 degrees.

3. Dimension the other values so the sketch matches Figure 7.19. Use the **Select** tool to drag the position of the dimensions to match the Figure.

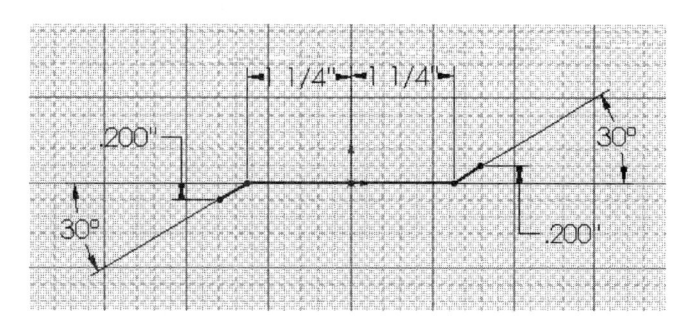

Figure 7.19.   Dimensions updated

4. A handy feature in the **Modify** dialog box is the stoplight button, which regenerates (redraws) the model with the new value. The stoplight button can be used to test the value in the dialog box. This is often helpful in determining what value looks reasonable as a design is modified. Using the **Select** tool, double click one of the "1 1/4" dimensions. Change the value to **2** and click the stoplight button in the **Modify** dialog box. The length of the horizontal line is increased in the sketch. If the green check button is clicked, the new value of 2 is used. If the red X button is clicked, the sketch reverts to the original value of $1\frac{1}{4}$. In this case, click the red X button.

5. After moving the dimensions, "shadows" of the dimension values might remain on the grid. Click the **Redraw** button (a paint bucket) on the **Standard** toolbar to update the screen.

7.2.5 Adding Fillets to a Sketch

The guard's bends are rounded. Fillets are rounded corners that will be used to round the sharp, angled bends on the guard's sketch.

1. While still in sketch mode, activate the **Fillet** tool by clicking the **Fillet** button (a rounded corner) in the **Sketch Tools** toolbar, or **Tools ⇒ Sketch Tools ⇒ Fillet**.
2. The **Sketch Fillet** dialog box appears, as shown in Figure 7.20. Type in **.3** to set the fillet radius to .3 inches.

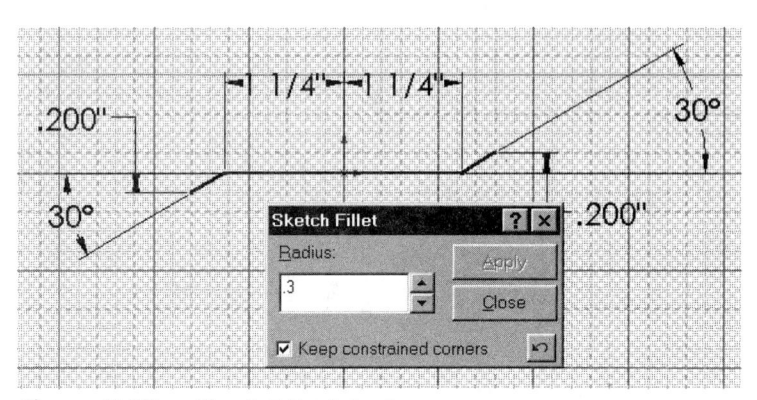

Figure 7.20. Sketch Fillet dialog box

3. Bring the cursor over the angled line on the right side. Notice that the line becomes highlighted in purple and an icon of a line appears next to the cursor. Click the line. It turns green, signifying that it is selected.
4. Click the horizontal line that touches the angled line you just selected. A fillet appears at the intersection of the two lines with a .3 dimension attached to it. A small cross also appears, showing the location of the center of the fillet's radius. Do not close the **Sketch Fillet** dialog box.
5. Repeat steps 3 and 4 to add a fillet to the left side of the guard. Notice that a dimension did not appear next to the second fillet. That is because the first dimension refers to both fillets.
6. Close the **Sketch Fillet** dialog box by clicking the **Close** button. With the **Select** tool, move the .3 dimension to an appropriate location. Your sketch should look similar to the one shown in Figure 7.21.

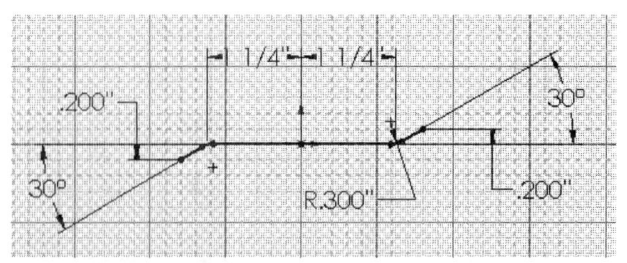

Figure 7.21. Fillets added

7.2.6 Extruding the Cross Section

Now that the sketch is both *Fully Defined* and properly dimensioned, it can be extruded into the third dimension. An extrusion "creates" material in the direction perpendicular to the plane of the sketch. For example, an extrusion of a circle would be a cylinder. In this case, the contour of the guard will be extruded to form the base feature of the guard. Right now the guard sketch shows line segments that form the edge of the guard. You will add thickness to these line segments and extrude the sketch to model the guard as a sheet metal part.

1. While still in sketch mode, click on the **Extruded Boss/Base** button in the **Features** toolbar, or select **Insert ⇒ Base ⇒ Extrude**. This is one of the only buttons in the **Features** toolbar on the left side of the screen that is highlighted. It causes a sketched section to be extruded perpendicular to the screen to create a three-dimensional part.

2. The view changes to an isometric view (the grid is at a 30-degree angle to the horizontal), and the *Extrude Thin Feature* dialog box appears, as shown in Figure 7.22. The dialog box may cover up the sketch. If so, move it to a new position by clicking on the blue bar at the top of the dialog box and holding the left mouse button down as you drag it to a new position on the screen. The guard is considered a *Thin Feature* because its sketch is a series of connected line segments rather than a closed polygon. Thin features have a depth into the drawing plane and a thickness, which can be thought of as the thickness of the lines of the sketch. In the *End Condition* tab, click on the small inverted triangle to pull down the *Type* menu. Select *Mid Plane*. This places the sketched lines at the midplane of the extrusion. In other words, the base feature will be extruded on both sides of the lines you have sketched to the front and back of **Plane1**. Set the *Depth* to *1*. This extrudes the sketched cross section 1/2 inch on both sides of the midplane.

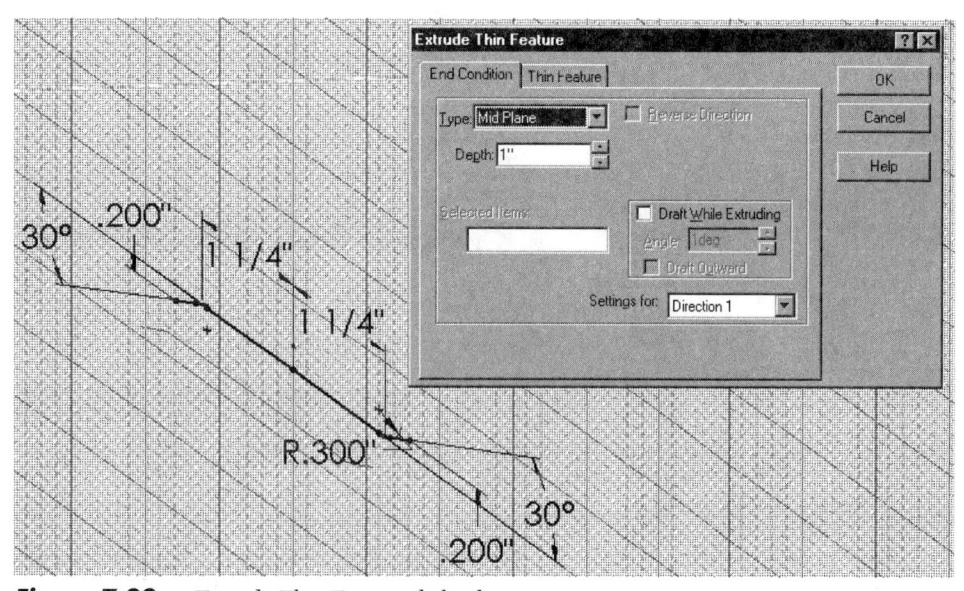

Figure 7.22. Extrude Thin Feature dialog box

3. In the *Thin Feature* tab, select *Mid-Plane* in the *Type* pull-down menu to indicate that the sketched lines will be at the midplane of the part's *thickness*. Type in the *Wall Thickness* of *.042*. This makes a thin part with its surfaces .021 inches above and below the lines that you have sketched. Click

OK in the dialog box to accept the settings. The screen should look similar to Figure 7.23. If it does not, you can remove the extrusion by clicking the **Undo** button (a curved arrow) in the **Standard** toolbar, or **Edit** ⇒ **Undo Base**. Then repeat steps 1–3.

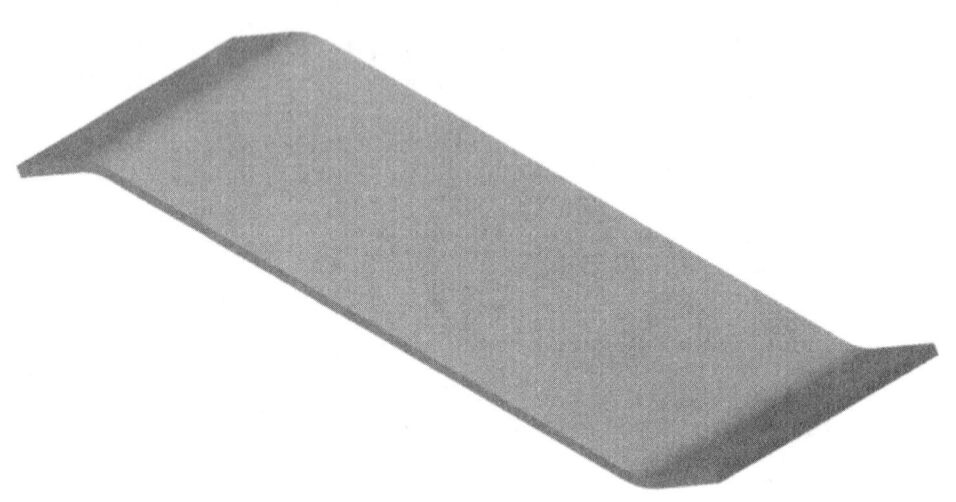

Figure 7.23.   Extruded guard

7.2.7   Viewing the Guard

Once you have extruded the sketch, several things in the SolidWorks environment change.

- The sketch mode is no longer active (no grid is shown on the screen).
- The FeatureManager design tree on the left-hand side of the screen shows a new feature, **Base-Extrude-Thin**, which is the base feature that you just created.

Now is a good time to use some of capabilities that SolidWorks has for orienting and viewing the part. Viewing options are available in the **View** menu and the **View** toolbar, shown in Figure 7.24. The **View** toolbar contains tools to orient, rescale, and display the part.

Figure 7.24.   View toolbar

1. Since you have made a three-dimensional representation of an object, you can view the part from any angle. Open the ***Orientation*** dialog box, shown in Figure 7.25, by selecting **Orientation** in the **View** menu, or by clicking the **View Orientation** button (a telescope) in the **View** toolbar. Move the ***Orientation*** dialog box to a convenient place on the screen by dragging its blue title bar. Be sure that it does not cover items in the FeatureManager design tree or the middle of the main Graphics Window. Click on the thumbtack on the top left of the dialog box to keep it open.

Section 7.2 Modeling the Guard   **103**

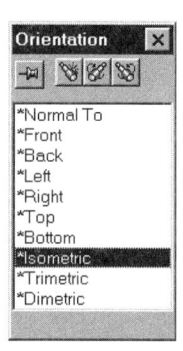

Figure 7.25. Orientation dialog box

2. The **Orientation** dialog box is used to quickly change the display of the part to a standard or custom view. In the dialog box, double click on **Dimetric** to see the part in dimetric view. Experiment with the other standard views by double clicking on the different views in the list, but avoid **Normal To** for now.

3. The **Normal To** orientation requires you to specify a planar face of the part that the view will be "normal to". To use **Normal To**, first double click **Isometric** so that the view looks like Figure 7.23. Next, select the guard's large planar face by clicking on it one time. The face will become highlighted. Now, double click on **Normal To** to show the part as viewed "normal to" this face. Return to the **Isometric** view.

4. The **View** toolbar also has tools that can be used to view the model at any angle, or to zoom in and out. Becoming accustomed to these tools will be helpful with orienting more complex parts. Click on the **Rotate View** button (two arrows forming a circle) and rotate the piece dynamically by moving the mouse while holding down the left mouse button. To translate the part on the screen, click on the **Pan** button (two crossed arrows) and drag while holding down the left mouse button. Use the **Zoom In/Out** button (magnifying glass with an arrow next to it) by clicking and dragging the cursor up to zoom in or down to zoom out. Return to the **Isometric** view. Click the **Zoom To Area** button (magnifying glass with a plus sign) to zoom in on a particular area of the Graphics Window. Note that the cursor changes to the same icon shown on the button. Move the cursor over the Graphics Window above the left end of the guard. Hold down the left mouse button as you drag the cursor downward and to the right. A blue box appears on the screen. When you release the mouse button, the view zooms to the area in the box. To get back to the standard view, click on the **Zoom To Fit** button (a magnifying glass with a box inside). Finally, you can zoom to a particular selected item using the **Zoom To Selection** button (a magnifying glass with two horizontal lines). To do this, return to the **Isometric** view and click on the **Select** button. Now, select the flat angled face of the guard at either the right or left end. The face should be highlighted. Next, click on the **Zoom To Selection** button to zoom in on this face. Return to the **Isometric** view and click the **Select** button so face is not highlighted.

5. Also in the **View** toolbar are the **Wireframe**, **Hidden In Gray**, **Hidden Lines Removed**, and **Shaded** buttons. Click each of these buttons to see the guard in wireframe, hidden lines shown in gray, hidden lines not shown at all, and shaded displays. The shaded view of the model is the

most realistic, but the other types of display can be helpful to see hidden faces of the part. The **Fast HLR/HLG** button activates a faster display for hidden lines removed or hidden lines in gray that is usually only needed for very complex parts.

7.2.8 Cutting a Hole in the Guard

The next feature to model is the hole at the center of the guard. This is done by extruding a circular solid cut through the guard at the origin.

1. Be sure that the ***Orientation*** dialog box is still open and double click ***Isometric*** to orient the view. Use the **Shaded** display.
2. With the **Select** tool, click on the large top face of the guard. The face becomes highlighted, indicating that it is selected.
3. Open a sketch on that face by clicking the **Sketch** button in the **Sketch** toolbar, or **Insert** ⇒ **Sketch**. A grid appears, indicating that you are in sketch mode. Notice that **Sketch2** has appeared in the FeatureManager design tree. Also note that the new grid is in the plane of the top face of the guard, which you just selected.
4. To orient the view so that it is facing the sketch plane, double click ***Normal To*** in the ***Orientation*** dialog box. This sets the view so that you look along a line normal to the surface just selected. Now, viewing the top face of the guard, you are ready to start sketching.
5. Select the **Circle** tool by clicking on the **Circle** button in the **Sketch Tools** toolbar, or **Tools** ⇒ **Sketch Entity** ⇒ **Circle**. Notice that the cursor has changed to a pencil with a circle below it.
6. Move the cursor to the origin. A small square appears next to the pencil cursor to indicate the cursor is over the origin. Click at the origin and hold the mouse button down to drag the radius of the circle outward from the origin. Release the mouse button when the circle is close to the size shown in Figure 7.26. The grid is not shown on this figure for clarity. If the circle is not quite the way you want it to be, it can be easily undone using **Undo Circle** in the **Edit** menu. The circle can then be redrawn using the **Circle** tool.

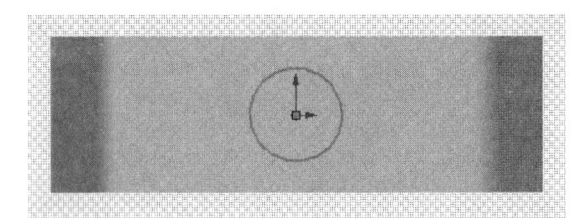

Figure 7.26.   Circle drawn

7. To dimension the circle, click the **Dimension** button in the **Sketch** toolbar, or select **Tools** ⇒ **Dimensions** ⇒ **Parallel**. Click somewhere on the arc of the circle. Click again to place the diameter dimension, as shown in Figure 7.27. The dimension appears with the lowercase Greek letter phi (ϕ) in front of the value. This indicates that the value refers to the diameter of the circle. A dimension beginning with **R** would indicate the radius of an arc or circle.
8. Use the **Select** tool to double click on the dimension. Set the value to **.4**. The screen should look like Figure 7.27.

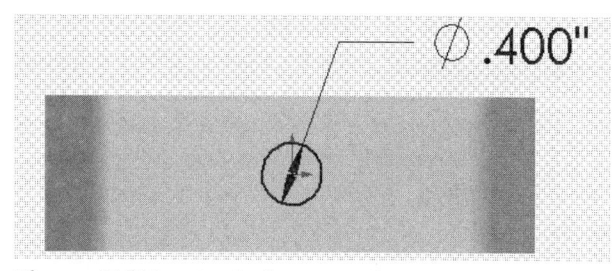

Figure 7.27. Circle dimensioned

9. Material can be removed by extruding a cut. Click on the **Extruded Cut** button (cube with a square hole) in the **Features** toolbar, or select **Insert** ⇒ **Cut** ⇒ **Extrude**.

10. The *Extrude Cut Feature* dialog box appears, which is shown in Figure 7.28. In the *Type* pull-down menu, select *Through All*. This option results in all of the material below the circle being cut away through all of the object. Other options include *Up To Next*, which makes a cut up to the next feature, and *Blind*, which makes a cut with a specified depth.

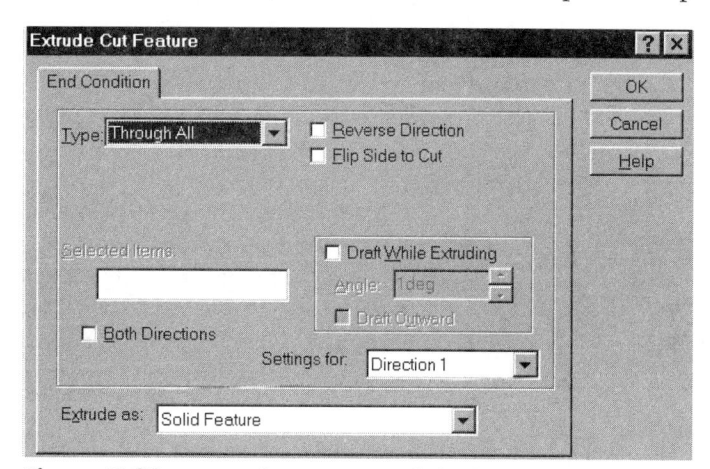

Figure 7.28. Extrude Cut Feature dialog box

11. Click *OK*. Using the **Rotate View** tool, rotate the part to see the newly created hole. In the *Isometric* view, the part should look like Figure 7.29.

Figure 7.29. Hole cut

7.2.9 Creating a Round at the Corners

Rounds on each corner are the last feature to add to the model of the guard. There are four corners that need to be rounded. This can be done one corner at a time or as a group. In this case, the rounds will be added one at a time. To do this, a feature called a **Fillet** will be created. When the bends of the guard were rounded, a fillet was created in the sketch mode. Here, the fillet will be applied as a feature rather than as a sketch entity.

1. Start in the ***Isometric*** view using the **Hidden Lines Removed** display. Zoom in to the corner closest to you by using the **Zoom to Area** button in the **View** toolbar. To do this, drag the **Zoom to Area** tool from the hole diagonally to the lower right corner to create a rectangular box. The result should look similar to Figure 7.30. If you zoom in too close, use either the **Zoom to Fit** or **Zoom In/Out** tool to change the viewing area.

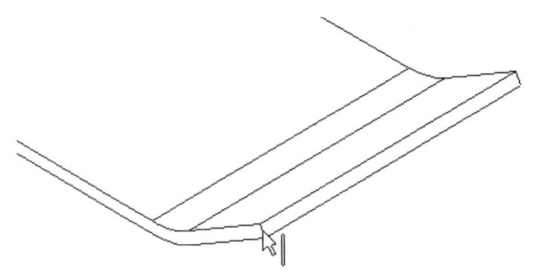

Figure 7.30. Cursor on corner

2. All of the features created to this point started out as sketches. In this case, the fillet feature will modify the existing features without using a sketch. Using the **Select** tool, bring the cursor over the edge at the lower right corner of the guard, as shown in Figure 7.30. This is the edge that will be rounded. Notice that the cursor changes to a curvy plane, a vertical line, or a square, showing that it is over a surface, a line, or a point, respectively. Select the line that forms the edge to be rounded by clicking on it.

3. To round the edge, click on the **Fillet** button (a cube with one edge rounded) in the **Features** toolbar or **Insert ⇒ Features ⇒ Fillet/Round**. The *Fillet Feature* dialog box appears, as shown in Figure 7.31. The selected edge, ***Edge <1>,*** is shown in the ***Items to Fillet*** list. If it is not there, you can select it while the dialog box is open. Set the ***Radius*** to **.35** and click **OK**.

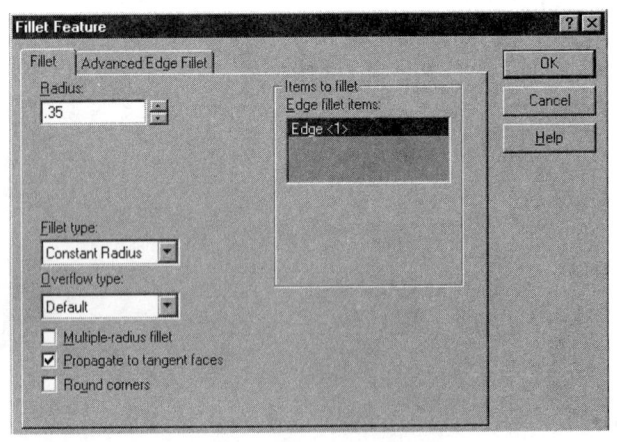

Figure 7.31. Fillet Feature dialog box

4. Click on **Zoom To Fit**, set the display to **Shaded**, and use **Rotate View** to see the filleted edge. The part should look similar to Figure 7.32.

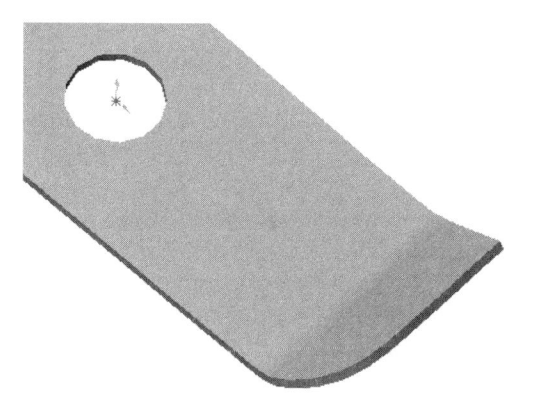

Figure 7.32.   Corner rounded

5. Set the display to **Hidden Lines Removed**. Add **.35** fillets to the other three corners by rotating the part, zooming in, selecting the edge, and using the **Fillet** tool. Note that SolidWorks assumes that the new fillets have the same radius as the previous fillets. Return to the **Isometric** view and click on **Shaded** after rounding all of the corners.

Congratulations! You have completed the model of the guard. The part should look similar to the one shown in Figure 7.5 at the beginning of this section. Click the **Save** button in the **Standard** toolbar, or **File** ⇒ **Save**. The name of the file and where it will be saved are specified in the *Save As* dialog box, as shown in Figure 7.33. Type in *guard* for the *File name*. The type of file, .sldprt, is automatically added to the filename that you specify. Navigate through the file system by clicking in the *Save in* field. Typically, you will save the part either to the hard drive (C:\) or to your own floppy disk (A:\). Click *Save* after specifying both the filename and folder.

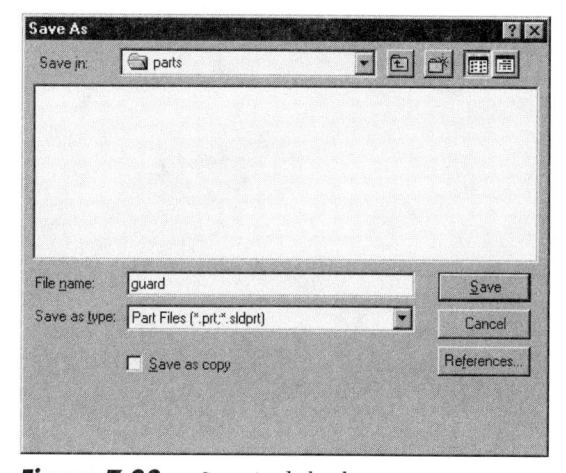

Figure 7.33.   Save As dialog box

It is also possible to print the image of the guard that appers in the Graphics Window. To do this, click on the **Print** button in the **Standard** toolbar (an icon of a printer near the left end of the toolbar), or **File ⇒ Print**. Printing is somewhat computer–printer system specific, so you may need to get more detailed printing instructions from your computer system administrator.

Close the guard using **File ⇒ Close**.

PROFESSIONAL SUCCESS; ADDING HOLES

The hole in the guard was modeled by sketching a circle and then using an **Extruded Cut** to remove the material. Holes can also be modeled without using the sketch mode. To do this, select the point on the surface where the hole is to be created. Then, use the **Simple Hole** or **Hole Wizard** tools, found in the **Features** toolbar, or in the **Insert ⇒ Features ⇒ Hole** menu.

The **Simple Hole** feature creates a sketch of a circle on the selected surface. Using the dialog box that appears, the diameter is specified and the material is removed using a **Blind**, **Through All**, or other method of cut. After the hole is cut, right click **Hole1** in the FeatureManager design tree and select **Edit Sketch**.

Then, add dimensions to position the hole's centerpoint using the **Dimension** tool. Finish by clicking the **Sketch** toolbar button.

The **Hole Wizard** can create more complex holes, such as countersunk, counterbored, and tapped holes. After selecting the point on the surface where the hole is to be created, activate the **Hole Wizard**. Select the hole type and modify its parameters in the dialog box. Click *Next* and locate the hole with the **Dimension** tool. Modify the dimensions to the desired values and click *Finish*. Using the **Hole Wizard** can save time in the design process, especially when many special holes are needed in a model.

7.3 MODELING THE ARM

The next component to model is the arm, which would be fabricated from stainless steel sheet metal of uniform thickness. One end of the arm is inserted into the handle while the other end holds the blade. The pizza cutter has two arms. Since the arms are identical, you will model one arm and use it twice in the assembly procedure outlined in Chapter 9. The finished arm is shown in Figure 7.34. The arm is modeled much as the guard was modeled, although a few new concepts will be introduced.

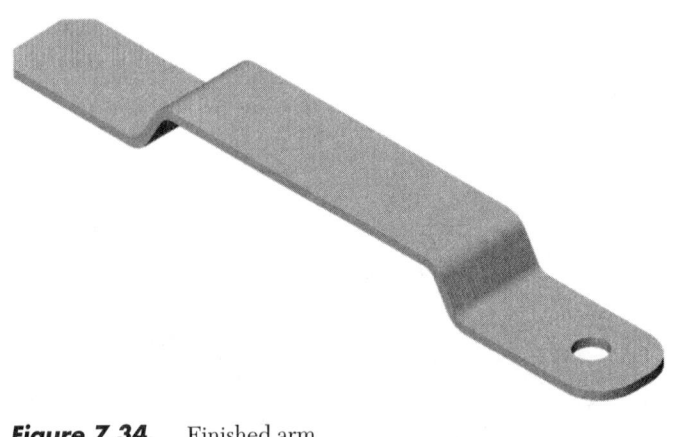

Figure 7.34.   Finished arm

7.3.1 Creating a New Part and a New Sketch

1. The arm will require a new part file. Select **File** ⇒ **New**, or click the **New** button in the **Standard** toolbar. Be sure **Part** is highlighted in the dialog box and click **OK**.

2. Click the **Grid** button to open the **Document Properties** dialog box. Set up the **Grid/Snap** and **Units** as in Figure 7.8 and click **OK**.

3. Open a new sketch by clicking the **Sketch** button from the **Sketch** toolbar, or select **Insert** ⇒ **Sketch**. The sketching grid appears. The first sketch for a new part always defaults to **Plane1** as the sketching plane. To see this, click **Plane1** in the FeatureManager design tree on the left-hand side of the screen. A rectangle appears which signifies that you are facing **Plane1**. The other two planes appear as lines, since you are viewing their edges.

7.3.2 Sketching the Lines

The initial sketch of the arm is similar to that of the guard. You will first draw several lines which indicate the shape of the arm's edge. Then, the sketch will be extruded into the third dimension as a thin feature. This creates the base feature for the arm.

1. In the sketch mode, use the **Line** tool (the **Line** button in the **Sketch Tools** toolbar) to draw a horizontal line through the origin beginning to its left and ending to its right. To do this, click the left mouse button to the left of the origin and drag it to the right past it. You should see an "H" next to the cursor as you draw a horizontal line. Be sure that the line passes through the origin. If it does not, or if the line is not horizontal, delete it. This can be done by selecting it (**Select** button in the **Sketch** toolbar) and then hitting the delete key. Alternatively, you could undo the line using **Undo Line** in the **Edit** menu, or by clicking the **Undo** button in the **Standard** toolbar.

2. With the **Line** tool still active, draw an angled line which starts from the right endpoint of the horizontal line you just drew and continues down and to the right. This time you should *not* see an "H" next to the cursor.

3. From the right endpoint of the angled line, draw a horizontal line that extends to the right.

4. Repeat steps 2 and 3 to make two lines on the left side of origin. The sketch should be similar to the one in Figure 7.35. The origin in the sketch should be on the upper horizontal line.

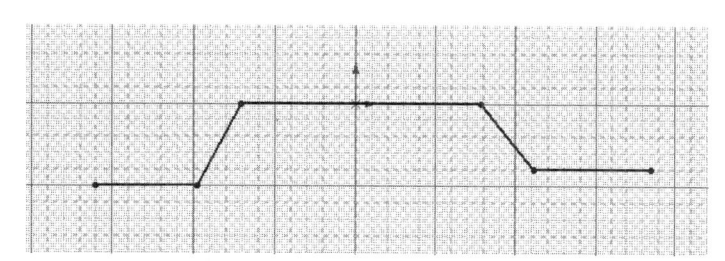

Figure 7.35. Lines sketched

7.3.3 Dimensioning the Sketch

1. Activate the **Dimension** tool by clicking on the **Dimension** button in the **Sketch** toolbar, or by selecting **Tools** ⇒ **Dimensions** ⇒ **Parallel**.
2. Dimension the middle horizontal line by clicking on the line itself. This is an alternative to clicking on both endpoints of the line. When the blue dimension box appears, place it above the line. Right now, the value of the dimension does not matter.
3. With the **Dimension** tool still active, dimension the two lower horizontal lines the same way you dimensioned the first line. Use Figure 7.36 as a guide.
4. Add a vertical dimension to the angled line on the right side of the origin by clicking on its endpoints. Remember that there are several ways in which this dimension can be placed, so be sure that you have chosen the vertical distance between the two endpoints as the dimension, and not the horizontal or aligned dimension.
5. To add the vertical dimension to the left angled line, first, click on the leftmost horizontal line segment. Next, click on the upper horizontal line segment and place the dimension. By choosing these line segments instead of the endpoints of the angled line, SolidWorks automatically assumes that this is a vertical dimension.
6. Next, you need to add a dimension that defines the angle of the angled line on the right side. With the **Dimension** tool active, click on the angled line to the right of the origin, then click on the upper horizontal line. The position of the dimension box is important, so look at Figure 7.36 before placing the dimension.
7. In the same manner, dimension the angle of the angled line on the left side. Except for the values of dimensions, the sketch should look similar to Figure 7.36. If necessary, delete a dimension by selecting it with the **Select** tool and hitting the delete key.

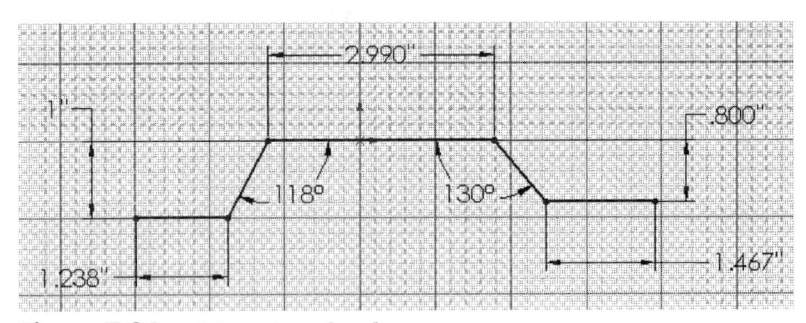

Figure 7.36. Dimensions placed

7.3.4 Adding a Relation

The Status bar at the bottom of the SolidWorks window indicates that the sketch is still *Under Defined*, even though all of the line segments have been dimensioned. The position of all five lines relative to the origin has not been defined. In other words, the entire sketched section needs to be located horizontally with respect to the origin. To add this constraint, you could dimension the distance from the origin to any point on the sketched section. Instead, a relation will be added that keeps the origin at the midpoint of the upper horizontal line.

1. Click the **Add Relation** button (perpendicular lines) in the **Sketch Relations** toolbar, or **Tools** ⇒ **Relations** ⇒ **Add**. The *Add Geometric Relations* dialog box appears. This dialog box permits the definition of relationships for entities in the sketch.
2. With the dialog box open, click on the origin point at the intersection of the tails of the perpendicular red arrows. Notice that the cursor has an asterisk next to it when it is near the origin. You may need to move the dialog box in order to see it. Once the origin is selected, *Point1@Origin* appears in the *Selected Entities* list. Since you want to orient the upper horizontal line with respect to the origin, click on the upper horizontal line to add it to the *Selected Entities* list. *Line1* appears in the *Selected Entities* list.
3. Select the *Midpoint* relations button (if it is not already selected). By selecting this button, the origin moves to the midpoint of the selected line. The screen should look similar to that shown in Figure 7.37.

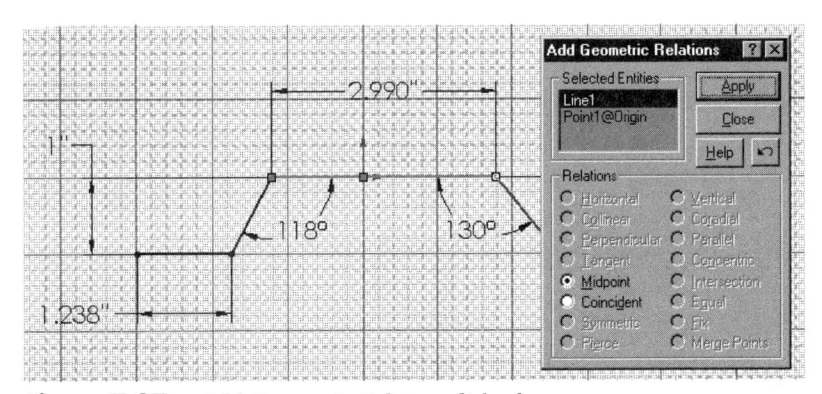

Figure 7.37. Add Geometric Relations dialog box

4. Click the *Apply* button. This horizontally shifts all of the lines in the sketched section to satisfy the constraint. The origin is now the midpoint of the upper horizontal line, regardless of its length or orientation. The sketch is now *Fully Defined*, as indicated in the Status Bar. Before closing the *Add Geometric Relations* dialog box, note the different types of relations that can be prescribed. Then *Close* the dialog box.
5. You may check all of the existing constraints by clicking the **Display/Delete Relations** button (reading glasses) in the **Sketch Relations** toolbar. The *Relations* tab shows the relations and constraints that each of the sketch entities in the sketch possess. Clicking on a sketch entity in the sketch shows the relations for that entity. Click on any of the horizontal lines and you will see that the *Horizontal* constraint is satisfied. This constraint occurred automatically when the "H" appeared as the line was sketched. Click on the upper horizontal line. The *Relations* tab indicates that this line has six relations (*Relation 1 of 6*). Clicking on the blue arrows at the bottom of the dialog box cycles through all six relations, highlighting the entities on the screen that are related. In addition to the origin being at the midpoint of the upper horizontal line, the other relations are that it is horizontal, coincident with the origin, a specified distance above the left horizontal line segment, and at an angle other than 90 degrees with respect to the angled lines (one relation for each angled line). The *Entities* tab shows the status of the selected sketch entity. Click *Close* to close the dialog box.

6. Using the **Select** tool, double click on the dimension that defines the length of the upper horizontal line. In the **Modify** dialog box, type in **1.6** and hit enter. The line length changes to the new dimension.

7. Do the same for the other dimensions using the values shown in Figure 7.38. Arrange the dimensions by clicking and dragging the dimension box with the **Select** tool. Redraw the screen using the **Redraw** button in the **Standard** toolbar to clean up the sketch. Click the **Zoom to Fit** button to make the sketch more easily readable.

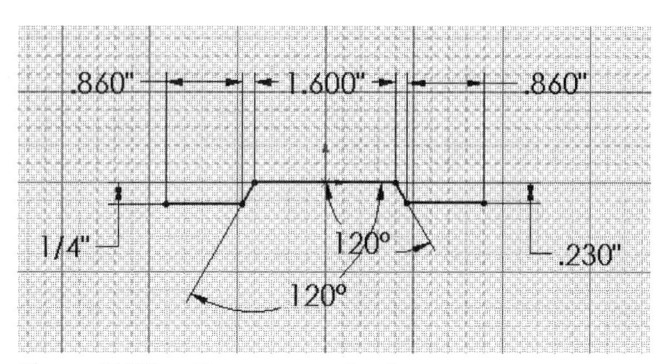

Figure 7.38. Dimensions updated

7.3.5 Extruding the Arm

Now that the sketch is *Fully Defined*, you can extrude the sketched section into the third dimension.

1. To extrude the sketched section, click **Extruded Boss/Base** in the **Features** toolbar. This changes the view to isometric and the **Extrude Thin Feature** dialog box appears. This is the same feature that was used to create the base feature of the guard. Again, the sketch in this case is a series of connected line segments rather than a closed polygon, indicating that the part is thin. This means that the part will have a uniform thickness along the sketched line segments, as well as a depth into the sketching plane.

2. In the **End Condition** tab, set the **Type** to **Mid Plane** so that the sketching plane is located at the mid plane of the extruded section. Set the **Depth** to **1/2**. This extrudes the arm 1/4 inches on each side of the sketch plane to obtain a total depth of 1/2 inches.

3. In the **Thin Feature** tab, set the **Type** to **Mid Plane** and the **Wall Thickness** to **.048**. This creates a thin sheet .048 inches thick, with the lines of the sketch at the center of the thickness. This thickness is a standard thickness for a type of stainless steel sheet metal referred to as 18 gauge.

4. The sharp bends at the intersections between the lines need to be rounded. This is because, in a real fabricated sheet metal part, bends are always rounded, not sharp. The rounded bends can easily be created using the **Auto Fillet** checkbox. Check the **Auto Fillet** box in the **Thin Feature** tab and set the **Fillet Radius** to **.075**. This produces the same results as inserting the fillets in the sketch of the guard.

5. Click **OK**. Examine the part by rotating and zooming to be sure that it looks like Figure 7.39. Change the view to isometric using the **View Orientation** button in the **View** toolbar to open the **Orientation** dialog box (if it is not already

open). Then double click **Isometric**. Note that the origin is at the center of the part's thickness and at the center of its width. (This is most evident when displaying the arm as **Wireframe** and rotating it, or when using the **Front** view.) Also, note that the bends in the arm are rounded as they would be if the part were formed out of sheet metal. If the arm does not look like the one in Figure 7.39, use **Edit ⇒ Undo Base** and start over with the extrusion.

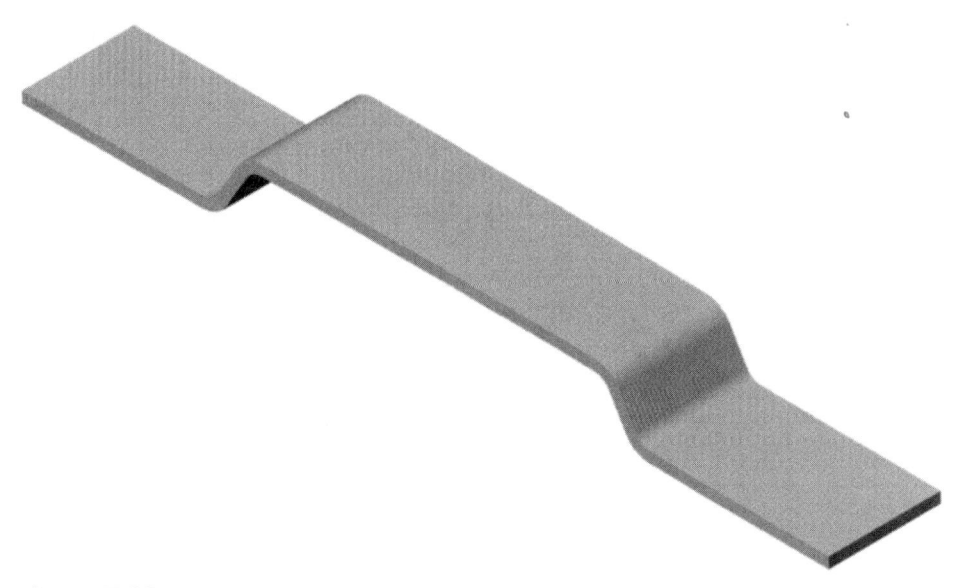

Figure 7.39.   Arm extruded

7.3.6   Rounding the End

The right end of the arm is rounded. This feature can be added to the base feature by using a fillet feature.

1. While in **Isometric** view with **Shaded** display, zoom in to the right end of the arm by using the **Zoom To Area** tool. Figure 7.40 shows the two corners to be rounded. Switch to the **Hidden Lines Removed** display.

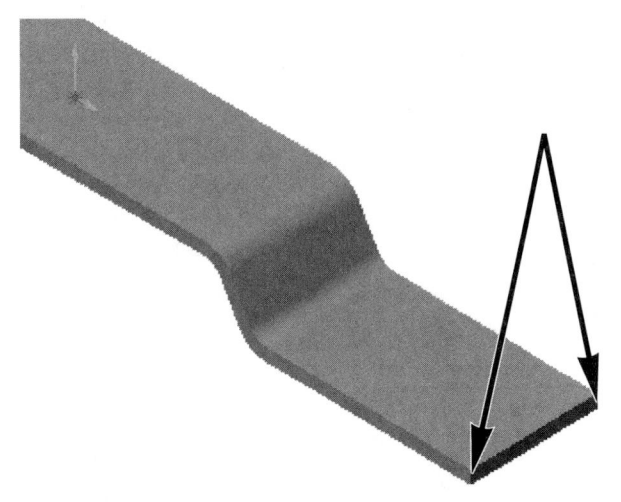

Figure 7.40.   Corners to be rounded

2. Several lines can be selected at one time. Click on the line forming the left corner edge with the **Select** tool to select it. Remember to look for the icon of a line next to the cursor when selecting. The selected line will turn to a blue dashed line, signifying that it is selected. To add another line to the selection, hold down the Control key on the keyboard while selecting the line forming the right corner edge. Both lines should be highlighted in blue.

3. Click the **Fillet** button in the **Features** toolbar, or **Insert** ⇒ **Features** ⇒ **Fillet/Round**. The *Fillet Feature* dialog box appears. Both lines are listed in the **Items to Fillet** as *Edge <1>* and *Edge <2>.* If both lines are not listed there, you do not need to close the dialog box. Just select the edge to be rounded, and it will be added to the list.

4. Set the *Radius* to *.2* and click **OK**. This creates two rounds, one on each edge selected. In **Shaded**, *Isometric* view, the part should look similar to that shown in Figure 7.41. Be sure that the rounded edges are the ones at the right end. If not, use **Edit** ⇒ **Undo Fillet** to start over on rounding the correct end.

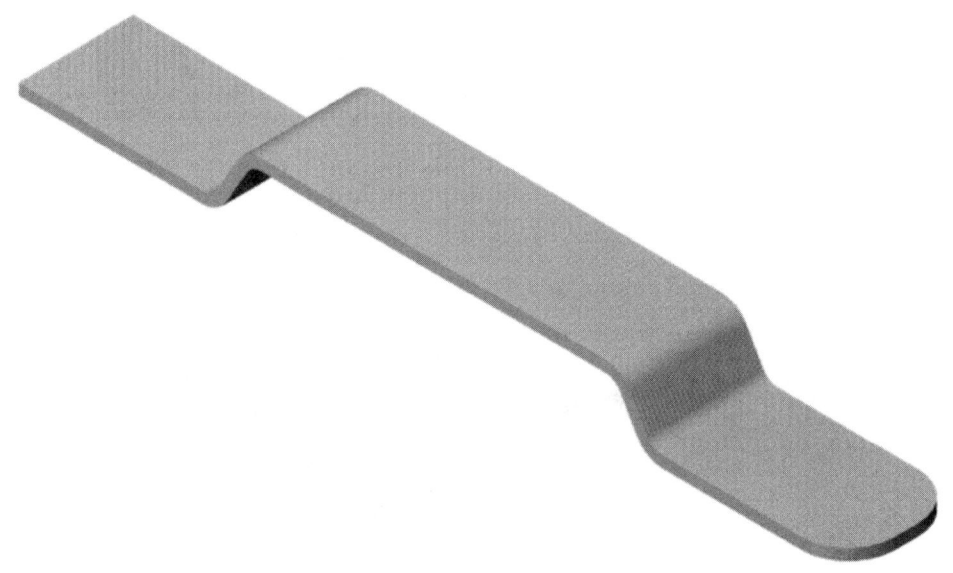

Figure 7.41.   Corners rounded

7.3.7   Adding Chamfers to the End of the Arm

The opposite end of the arm will have chamfers. A chamfer is an angled "cutoff" of a corner.

1. Open the *Orientation* dialog box, if it is not already open. Double click *Left* from the list of standard views. This shows the arm from the left side, which does not have the rounded corners. If necessary, click the **Hidden Lines Removed** button in the **View** toolbar to make the view a little clearer.

2. To understand what you are looking at, it is a good idea to rotate the part slightly in order to orient yourself. Click the **Rotate View** button to rotate the part slightly so it looks like that shown in Figure 7.42. Use the **Zoom In/Out** button in the **View** toolbar to zoom out. Be sure that you can still see the entire end of the arm.

3. To add the chamfer, click the **Chamfer** button (cube with one edge chamfered) in the **Features** toolbar, or **Insert** ⇒ **Features** ⇒ **Chamfer**. The *Chamfer Feature* dialog box appears. Move the dialog box so that it does not cover up the end of the arm.

Section 7.3 Modeling the Arm   **115**

4. Select both of the edges shown in Figure 7.42 by clicking each one. Each edge is added to the **Items to Chamfer** list in the dialog box. Note that it is not necessary to *control click* the second edge as was done for the round. This is because the **Chamfer Feature** dialog box was opened before the edges were selected. Once the feature is chosen, several edges can be selected. In the case of the fillet, the edges were selected before clicking the **Fillet** button. Prior to opening a feature dialog box, *control clicking* must be used to select more than one entity.

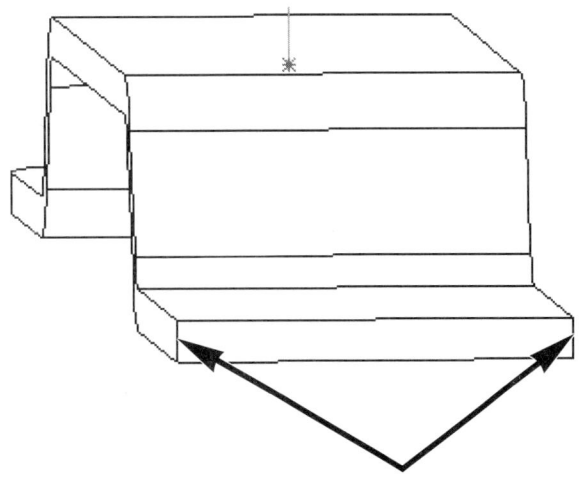

Figure 7.42.   Corners to be chamfered

5. With the **Chamfer Type** set to **Angle-Distance**, set the **Distance** to **.1** and the **Angle** to **45** degrees. This will cut off the edge at a 45-degree angle so that the side of the triangle is .1 inches long.

6. Click **OK** to finish the chamfers. This creates two chamfers, one on each edge you selected. Click the **Shaded** button in the **View** toolbar. In the isometric view, the part should look similar to that shown in Figure 7.43. Be sure that the chamfers are on the left end of the arm, as shown in the Figure.

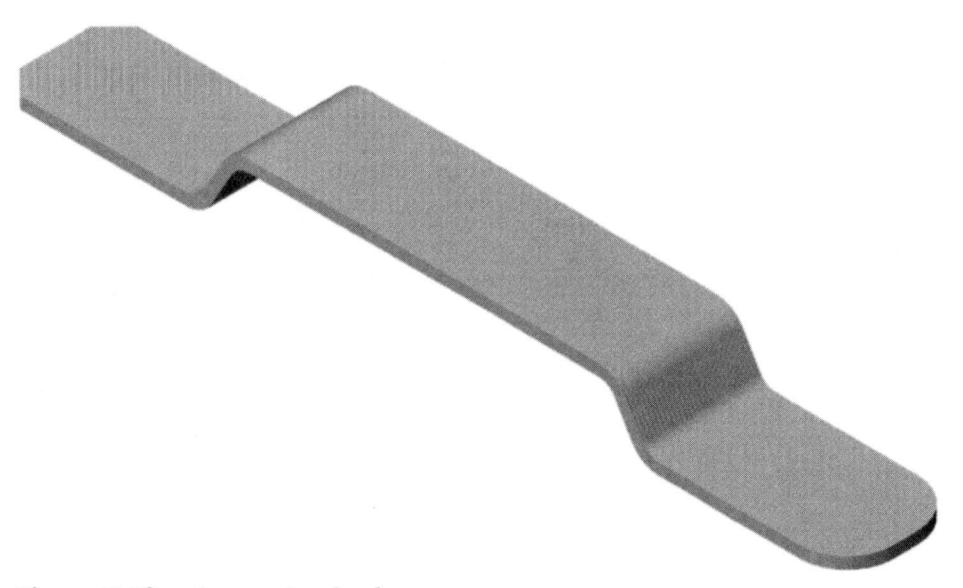

Figure 7.43.   Corners chamfered

7. To see the surfaces defined by the chamfer, click on **Chamfer1** in the FeatureManager design tree on the left side of the window. Both of the chamfers will be highlighted in green. Double clicking on **Chamfer1** in the FeatureManager design tree will display the chamfer's dimensions. If the dimensions are difficult to read, drag them to a better position. Use **Zoom to Fit**, if necessary. Double-clicking on **Fillet1** or **Base-Extrude-Thin** in the FeatureManager design tree will display dimensions of these features. The dimensions could be modified by double clicking on the values, but do not do this now. The dimensions can be displayed in any of the views in the **Orientation** dialog box. Activate the **Select** tool to hide the dimensions, and use **Redraw** to repaint the screen, if necessary.

7.3.8 Adding a Hole in the Arm

The last feature to add to the arm is the hole at the rounded end. This can be accomplished by extruding a circular cut through the arm.

1. Return to the *Isometric* view and click on the top face with the rounded edges at the right end of the arm using the **Select** tool. The face becomes highlighted. Open a sketch on the face by clicking the **Sketch** button in the **Sketch** toolbar. Double click *Normal To* in the **Orientation** dialog box so that the view is directed toward the selected face. The screen should look similar to that shown in Figure 7.44, although the guard might be perpendicular to that shown.

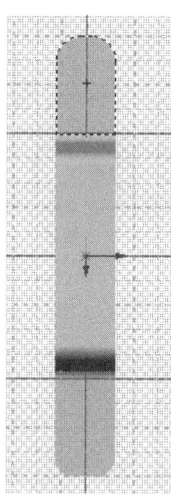

Figure 7.44. Face selected

2. Click the **Circle** button in the **Sketch Tools** toolbar, or **Tools** ⇒ **Sketch Entity** ⇒ **Circle**. The circle next to the cursor signifies that the **Circle** tool is active. Draw a circle near the end of the arm by clicking and dragging the cursor with the left mouse button. At this point, it does not matter what the diameter is or where the center is located. However, if the center of the circle is on an edge, delete the circle and redraw it.

3. Activate the **Dimension** tool by clicking the **Dimension** button in the **Sketch** toolbar, or **Tools** ⇒ **Dimensions** ⇒ **Parallel**. Dimension the diameter of the circle by clicking on the circle and then clicking again to place the dimension box.

4. The center of the circle must be located relative to the arm. In this sketch, the center of the circle will be dimensioned from existing features (the edges of the arm). With the **Dimension** tool active, click on the center point of the circle. Next, click on the left edge of the arm. Click again to place the dimension. The sketch should look similar to that shown in Figure 7.45. The dimension values may be different.

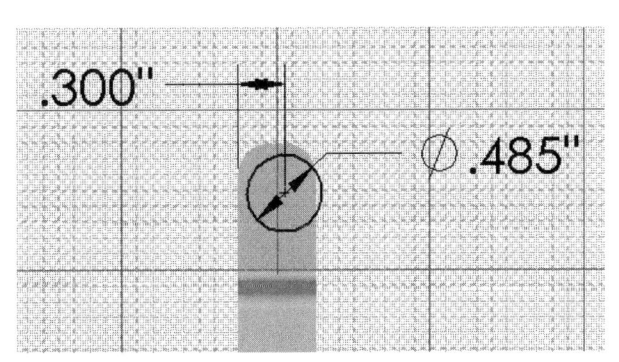

Figure 7.45. Circle with dimensions placed

5. The circle center point needs a second dimension to locate it with respect to the arm. The **Dimension** tool should still be active, so click on the center of the circle. Click on the edge of the arm that is between the two rounds at its rounded end. Before you click the line, be sure that an icon of a line appears next to the cursor. Zoom in, if you have trouble selecting the line. Click again to place the dimension.

6. Activate the **Select** tool to change the values of the dimensions to match the ones shown in Figure 7.46.

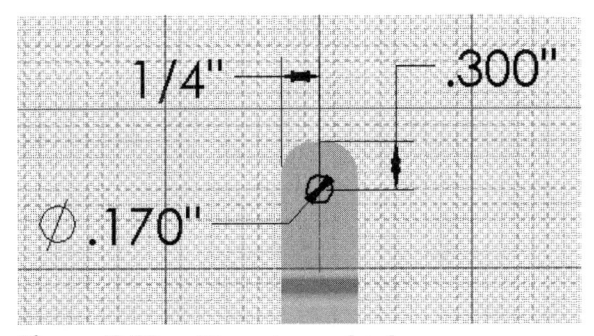

Figure 7.46. Dimensions updated

7. The circle will become a hole when it is extruded as a cut. Click **Extruded Cut** in the **Features** toolbar. In the *Extrude Cut Feature* dialog box set the *Type* to *Through All*, so that the cut goes all the way through the thickness of the arm. Click **OK** to finish the cut.

8. Rotate the part to be sure that the hole was created correctly. It should look like Figure 7.34.

Congratulations! This completes the arm. Save the part as *arm* by clicking **Save** in the **File** menu or by clicking the **Save** button in the **Standard** toolbar. Close the arm window.

7.4 MODELING THE BLADE

The third component in this chapter is the blade of the pizza cutter, shown in Figure 7.47. The blade could be modeled using an extruded circular section as the base feature, with a circular cut for the hole as an additional feature. But in this case, you will include the circular hole at the center of the blade in the base feature. This avoids adding the hole later as a separate feature. The chamfered edges are then added to the base feature.

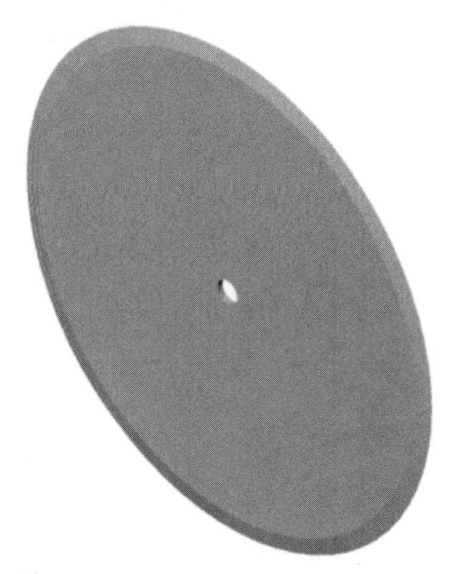

Figure 7.47. Finished blade

7.4.1 Sketching the Blade

The initial sketch for the blade consists of two concentric circles. The larger circle is the edge of the blade. The smaller circle is for the hole at the center of the blade.

1. Open a **New** part, set up the **Grid/Snap** and **Units** using the **Grid** button, and then open a new **Sketch**.
2. Activate the **Circle** tool and draw a small circle that is centered at the origin. Be sure that you see the square next to the cursor before you place the center of the circle, signifying that the center is coincident with the origin.
3. Draw a larger circle that is also centered at the origin. The sketch should look similar to that shown in Figure 7.48.

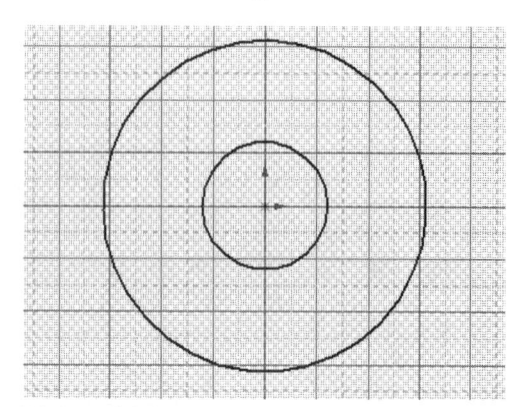

Figure 7.48. Circles drawn

4. **Dimension** the inner circle to have a diameter of **.170** by clicking on the arc of the circle and placing the dimension. Dimension the outer circle to have a diameter of **3 1/2**. Use the **Select** tool to modify the values. The sketch should look similar to that shown in Figure 7.49.

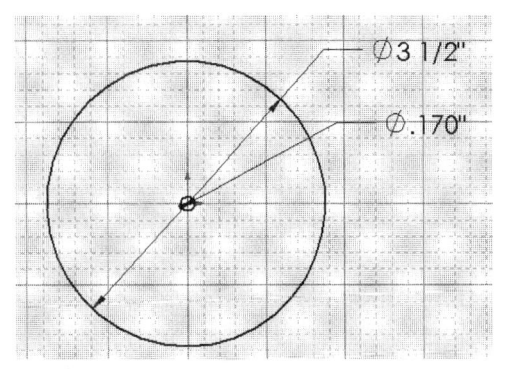

Figure 7.49. Dimensions placed and updated

7.4.2 Extruding the Sketch

The sketch is extruded perpendicular to the screen to create a cylinder with a hole through it. The extrusion depth is the thickness of the blade.

1. Open the **Extrude Feature** dialog box by clicking the **Extruded Boss/Base** button in the **Features** toolbar.
2. The two circles form a closed geometric shape. Since the shape is closed, it is not a thin feature having a wall thickness and a depth as the guard and arm did. Instead, it is a solid feature in which the sketched geometric shape is extruded perpendicular to the sketching plane. Thus, only the **End Condition** tab appears in the dialog box, and the **Extrude As** field is set to **Solid Feature**. Set the **Type** to **Blind** and the **Depth** to **1/2**. This extrudes the sketch 1/2 inches in one direction from the plane of the sketch.
3. Click **OK** to finish the extrusion. The extruded part should look similar to that shown in Figure 7.50. The final blade will be quite thin. However, larger objects are easier to work with, so a very thick blade was extruded. The thickness of the blade will be changed later.

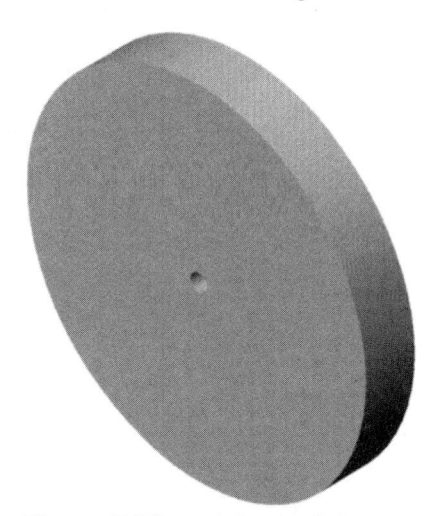

Figure 7.50. Blade extruded

7.4.3 Adding a Chamfer to Form the Edge of the Blade

1. To create a chamfer, the edge indicated in Figure 7.51 must be selected. This edge is opposite the face of the extrusion that lies in **Plane1**. To see where **Plane1** is, select it in the FeatureManager design tree on the left side of the screen. Its location is more easily seen from the right side. Double click *Right* in the **Orientation** dialog box. Set the view to **Hidden In Gray**. Click on the vertical edge that does *not* lie in **Plane1**. Return to the *Isometric* view and **Shaded**. The edge should be highlighted as shown in Figure 7.51.

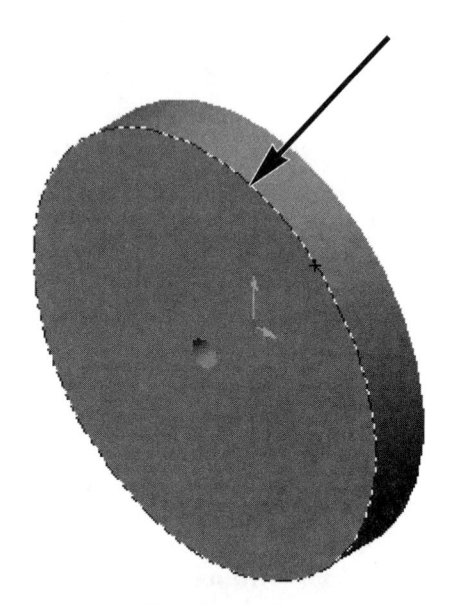

Figure 7.51. Selected edge

2. Add a chamfer to this edge by clicking the **Chamfer** button in the **Features** toolbar, or **Insert ⇒ Features ⇒ Chamfer**.

3. In the **Chamfer Feature** dialog box, be sure that **Edge <1>** is in the **Items to Chamfer** list.

4. Set the **Type** to **Distance-Distance**. This removes a triangular section with its sides defined by two distances. The arrow shown on the part while the **Chamfer Feature** dialog box is open points in the direction of **Distance 1**. Set **Distance 1** to **.02** and set **Distance 2** to **.12**.

5. Click **OK** to create a chamfer around the entire edge. Look at the part using **Hidden Lines Removed**, and be sure that the long side of the chamfer is on the flat face of the extrusion. Double click on **Chamfer1** in the FeatureManager design tree to show the chamfer dimensions as in Figure 7.52. You may need to move the dimensions using the **Select** tool in order to see them better. If the chamfer is not as desired, click on **Chamfer1** in the FeatureManager design tree to highlight it. Hit the delete key on the keyboard to delete the chamfer feature. Redo the feature, making sure the settings are the same as those stated previously.

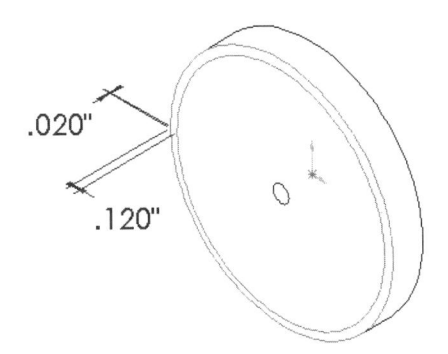

Figure 7.52. Chamfer dimensions shown

7.4.4 Mirroring a Feature

In many cases, a feature is symmetric about a plane of the part. Symmetric features are easily modeled by creating a mirror image of the feature. Mirroring both the extrusion and the chamfer will create a new set of features on the other side of **Plane1**. This will form the other half of the blade.

1. The two features that will be mirrored, the cylindrical base feature and the chamfer, must both be selected. To select these two features, click on **Base-Extrude** in the FeatureManager design tree. The extrusion turns green to signify that it is selected. Then *control-click* **Chamfer1** in the FeatureManager design tree, adding it to the selection.

2. Click **Insert** ⇒ **Pattern/Mirror** ⇒ **Mirror Feature** in the Menu bar. The *Mirror Pattern Feature* dialog box appears.

3. The two features, **Base-Extrude** and **Chamfer1,** should appear in the *Features to mirror* list. These features will be mirrored about **Plane1**. Be sure that the *Mirror Plane* field is highlighted in pink. Click on **Plane1** in the FeatureManager design tree, adding it to the *Mirror Plane* list. **Plane1** and the mirrored features now appear, as shown in Figure 7.53.

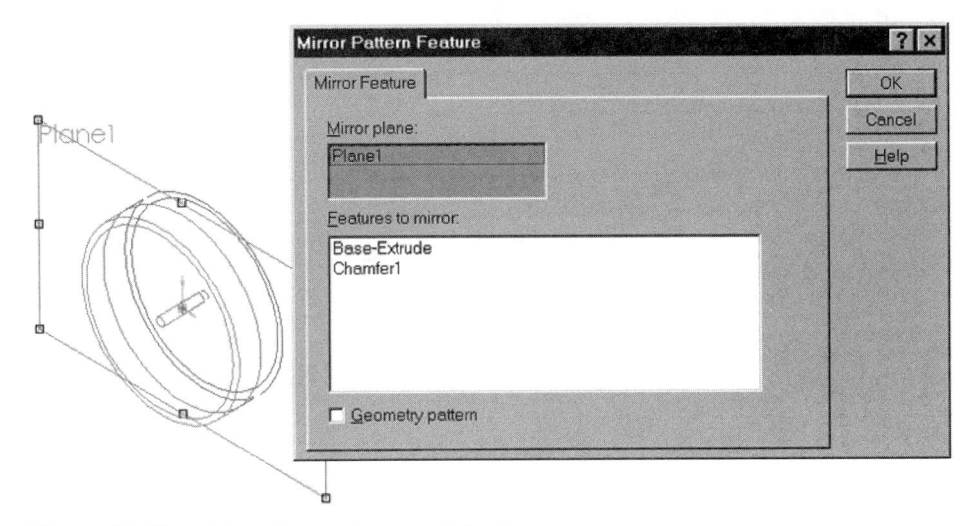

Figure 7.53. Mirror Pattern Feature dialog box

4. Click **OK** to accept the settings. Rotate the part to be sure both the extrusion and the chamfer were copied. If there is a problem with the mirror operation, select **Mirror1** in the FeatureManager design tree, delete it, and redo the previous steps.

7.4.5 Changing the Definition of the Extrusion

You may now go back and change the extrusion depth for the blade. Since the chamfer and the mirror operation were done *after* the extrusion, they will be updated to reflect the new blade thickness.

1. In the FeatureManager design tree, *right click* **Base-Extrude**. Using the right mouse button activates a menu. Choose **Edit Definition** in the menu. The ***Extrude Feature*** dialog box appears, with the settings that were originally specified for the extrusion of the blade's base feature.

2. Change the ***Depth*** to **.02**. This changes the thickness of the blade to exactly the width of the chamfer. Leave the other settings as they were.

3. Click **OK**. SolidWorks rebuilds all of the part's features, including the mirror operation with the new depth. The final blade should look like Figure 7.47. If an exclamation point appears next to **Chamfer1** in the FeatureManager design tree, select **Edit Definition**. Be sure that the settings match the ones described above.

Congratulations! This completes the blade. Save the part as ***blade*** and close the window. If you are going to quit now, use **File ⇒ Exit**.

PROFESSIONAL SUCCESS: PARENT–CHILD RELATIONSHIPS

The concept of parent–child relationships is important in solids modeling. A child is a feature that depends on another feature, namely the parent feature. For example, the chamfer created on the blade is a child of the base extrusion for the blade, since the chamfer cannot exist without the base extrusion. Therefore, the base extrusion is a parent of the chamfer. A parent can have multiple children (if several different features are dependent upon it), and a child can have multiple parents (if it is dependent on several different features).

To see the parent–child relationships in SolidWorks, *right click* a feature in the FeatureManager design tree and select Parent/Child. The parents and children of that feature are shown. These relationships are important because features within SolidWorks maintain their dependencies if the parent features are changed. This was demonstrated when the blade was modeled. When the thickness of the blade was changed, the chamfers (children) remained on the edge of the blade base feature (parent), as defined earlier on the thicker blade. It is important to keep track of parent–child relationships, since altering or deleting a parent feature might adversely affect the children of that feature.

KEY TERMS

Chamfer	Extrude	Mirror
Constraints	Feature	Relation
Cross section	Feature-based modeling	Sketch
Cut	FeatureManager design tree	Solid feature
Datum planes	Fillet	Thin feature

Modeling Problems

1. Model the guard by making both ends bend 30 degrees in the same direction, rather than in opposite directions.
2. Model the guard so that it is extruded only to one side of **Plane1** using **Blind** instead of **Mid Plane** in the **Extrude Thin** dialog box. Make the radius of the round *.15* instead of *.35*.
3. Model the arm by using a **Blind** depth extrusion instead of a **Mid Plane** extrusion so that it is extruded to one side of **Plane1**. Change the chamfer to a *.15* fillet.
4. Create the blade by first creating a *3 1/2* solid disk with a **Blind** depth of *1/2*. Add the *.170* hole at the center. Then mirror both the extrusion and the hole. Create the chamfers using **Angle-Distance**. The angle is that necessary to create the same chamfer as described in this chapter. Note that the chamfer distance is in the direction of the arrow shown when the **Chamfer Feature** dialog box is open. Modify the definition of the extrusion to provide the correct blade thickness.
5. Model the blade by creating a *3 1/2* disk with a *.170* hole, but for the **Extrude Feature**, use **Mid Plane** with the actual final blade thickness of .040 inches. Finally, add a chamfer on both sides of **Plane1** to form the cutting edge of the blade.

8
Modeling Parts in SolidWorks: Revolves

OVERVIEW

In this chapter, you will learn how to create three-dimensional solids models by revolving a two-dimensional cross-section. Revolving is ideal for creating axisymmetric models, such as the pizza cutter's cap, handle, and rivet. Modeling a revolved solid begins in the sketch mode. But, unlike an extrusion, a centerline is added to the sketch to indicate the axis about which the cross section is revolved. After the revolved base feature is created, additional features such as fillets, holes, and other cuts can be added, including features that are repeated in a pattern. In addition, parts can be colored to look much like the actual part.

8.1 MODELING THE CAP

The cap covers the end of the handle, where the two arms of the pizza cutter are inserted into the handle. The cap, shown in Figure 8.1, has a uniform thickness and is axisymmetric. The cap can be modeled as a thin feature by revolving a single sketch of a cross section around a centerline.

8.1.1 Sketching the Cap

You will begin by sketching the thin section that will be revolved.

SECTIONS
- 8.1   Modeling the Cap
- 8.2   Modeling the Handle
- 8.3   Modeling the Rivet

OBJECTIVES

After working through this chapter, you will be able to

- Sketch centerlines
- Revolve a sketch to model a solid body
- Revolve a sketched geometry to add a revolved feature to a base feature
- Use the Fillet, Rectangle, and Trim sketch tools
- Create an axis from the intersection of two planes
- Create multiple identical features
- Color a part
- Sketch arcs
- Change detailing settings to make a sketch clearer

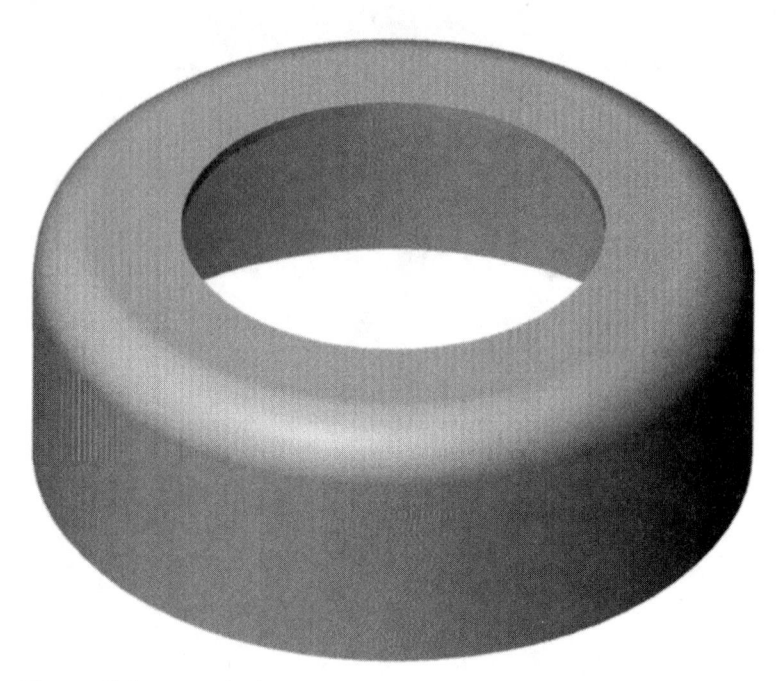

Figure 8.1. Finished cap

1. Open a new sketch in a new part using **File** ⇒ **New** and selecting the <u>**Sketch**</u> button, or **Insert** ⇒ **Sketch**. Set up the grid and units using the <u>**Grid**</u> button as in the previous chapter.

2. With the **Line** tool, draw a horizontal line that starts to the right of the origin and extends further to the right as shown in Figure 8.2. Remember to look for an "H" next to the cursor when drawing horizontal lines.

3. The **Line** tool should still be active. Draw a vertical line that starts at the right endpoint of the horizontal line and continues downward. Look for two things when drawing this line. When you start the line, a square should appear next to the cursor. This signifies that the line will start at the endpoint of the horizontal line. As you draw the line, look for a "V", indicating that a vertical line is being drawn. If you need to correct a mistake, use <u>**Undo**</u> or **Edit** ⇒ **Undo Line** to remove the line. Then draw the line again. The sketch should look like Figure 8.2.

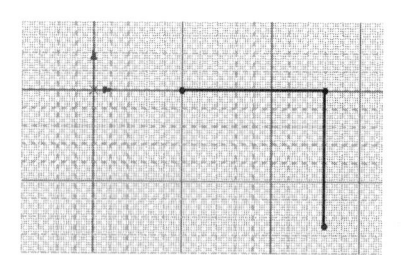

Figure 8.2. Lines sketched

4. Activate the **Dimension** tool. Dimension the distance from the origin to the left endpoint of the horizontal line to **.3**. To do this, click on each point and then click again to place the dimension. Use the **Select** tool to modify the value by double clicking on the dimension.
5. Dimension the horizontal line to **.2** by clicking on the line, placing the dimension, and then modifying the value.
6. In a similar manner, dimension the vertical line to **3/8**. The sketch is now fully defined. The screen should look similar to that shown in Figure 8.3. Use the **Select** tool to move the dimensions to look like the figure. It may help to use the **Zoom to Fit** button in order to see a close-up view of the sketch.

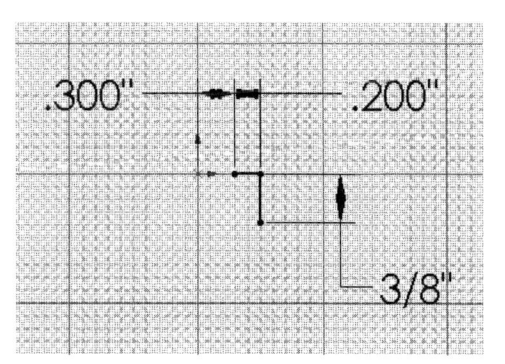

Figure 8.3. Dimensions placed and updated

7. Now, round the corner between the two line segments by clicking on the **Fillet** button in the **Sketch Tools** toolbar, or **Tools** ⇒ **Sketch Tools** ⇒ **Fillet**.
8. The *Sketch Fillet* dialog box appears. Type in *.075* as the *Radius*.
9. With the dialog box still open, click the horizontal line, and then click the vertical line. The corner is rounded, and a dimension is automatically placed on the sketch.
10. Click **Close** to close the **Sketch Fillet** dialog box. Using the **Select** tool, move the fillet dimension to a location where it can be seen clearly.

8.1.2 Adding a Centerline

Centerlines are frequently used in SolidWorks. Centerlines are *construction geometry* and aid in creating a sketch of a part, but are not the part itself. Centerlines are used in mirroring objects, revolving sketches, and constraining endpoints to be collinear. You will create a vertical centerline through the origin that will serve as an axis of revolution for the shape that has been sketched.

1. Activate the **Centerline** tool by clicking the **Centerline** button in the **Sketch Tools** toolbar, or **Tools** ⇒ **Sketch Entity** ⇒ **Centerline**.
2. Centerlines are drawn just like any other line. With the **Centerline** tool, click on the origin and drag the centerline upward. Let go of the mouse button when there is a "V" next to the cursor. A centerline extends infinitely (even though it has a finite length on the sketch), so you do not need to dimension it. Be sure that the centerline is vertical and begins at the origin. The sketch should look similar to that shown in Figure 8.4.

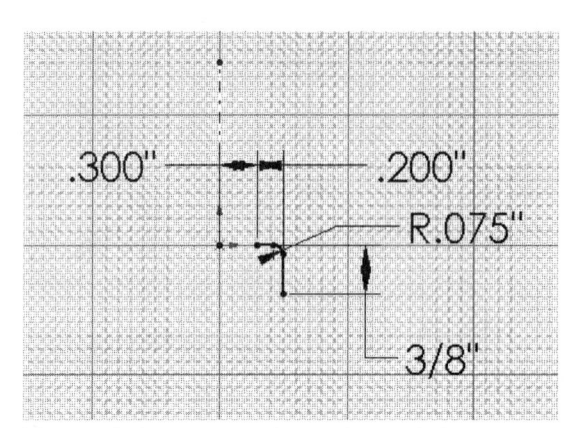

Figure 8.4. Fillet and centerline added

8.1.3 Revolving the Sketch

You are ready to revolve the sketch around the centerline in order to model a three-dimensional, axisymmetric part.

1. Now that a centerline exists in the sketch, the **Revolved Boss/Base** button (half of a ring) in the **Features** toolbar is active. Click this button, or select **Insert ⇒ Base ⇒ Revolve**.

2. The dialog box shown in Figure 8.5 will appear, since the sketch is not a closed outline. The cap is a thin feature, so the sketch outline should not be closed. If you answered yes to the question in the dialog box, SolidWorks would automatically draw a line between the left end of the horizontal line and the bottom end of the vertical line, to create a closed triangle with a rounded corner. Then, SolidWorks would revolve this triangle around the axis. Instead, you should revolve the thin shape that was sketched, so click **No** to continue.

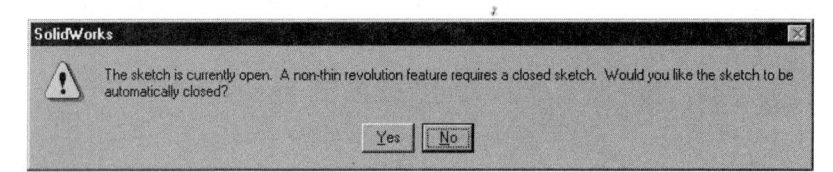

Figure 8.5. Open sketch dialog box

3. The **Revolve Thin Feature** dialog box appears. In the **Revolve Feature** tab, set the **Type** to **One-Direction**. This causes the revolved section to extend in one direction from the sketch plane. Using **Two-Direction** revolves the sketch a specified angle on one side of the sketch plane and a different specified angle on the opposite side of the sketch plane. **Mid-Plane** revolves the sketch a specified angle, but places the sketch plane so that it bisects the angle.

4. Set the **Angle** to **360** degrees. This revolves the sketch completely around the centerline, which creates a body of revolution.

5. In the **Thin Feature** tab, set the **Type** to **Mid-Plane**. Set the **Wall Thickness** to **.02**. This creates a thin feature .02 inches thick, with the sketch lines at the middle, or mid plane, of the thickness.

6. Click **OK** to accept the settings. Rotate the part using **Rotate View** to see all sides of the cap. Notice that it is symmetric about the centerline and has a uniform wall thickness of .02 inches. To return to the isometric view, use **View ⇒ Orientation**, or the **View Orientation** toolbar button, to activate the **Orientation** dialog box and *double click* on **Isometric**.

Congratulations! This completes the cap. The part should look like Figure 8.1. **Save** the part as *cap*. Close the cap window using **File ⇒ Close** or by clicking the X in its upper right corner (not the SolidWorks window).

Revolving a section can transform a simple sketch into a relatively complex part. The cap could have been modeled using an extrusion of a circle to create a cylinder, followed by various cuts and fillets to remove material. But this would have required several features. By using a revolved section, the cap was modeled using just one feature. In fact, the blade could have been made with a revolve in only one step using the sketch shown in Figure 8.6. When modeling in three dimensions, there are usually several different ways of creating parts. Some are more efficient than others. As a result, it is helpful to visualize the part and think about the options available before modeling it.

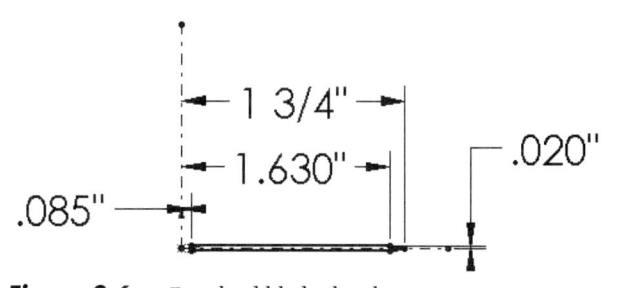

Figure 8.6. Revolved blade sketch

PROFESSIONAL SUCCESS: WHAT IS THE BEST APPROACH TO MODEL A PART?

Many different approaches to model the same part are possible. Which is the best approach? The answer to this question depends on the skill of the CAD user, the complexity of the part, the potential for redesign of the part, the design intent, and the preferences of the user. Unskilled users may prefer to "carve up" a solid block of material by using rounds, cuts, and chamfers, whereas skilled users may extrude or revolve a cross section that already includes many of these features. Simple parts can often be modeled with only a few strategic features, but complex parts may require hundreds of features.

For a simple part, the order in which features are modeled is unlikely to be critical. For a complex part, it may be necessary to model one feature before another one can be created. If redesign of the part is likely, several features may be separately added to the model, so that each could be individually removed or modified. If redesign is unlikely, these features could be combined. Design intent can play a key role in both how a feature is created and which features are modeled first. The parent-child relationships, how features are dimensioned with respect to other features, and what features are likely to be modified all play a role. Finally, some users prefer to model using the fewest features possible by extruding or revolving very detailed sketches. Others prefer to mimic the operations that a machinist might use in making a part. They begin with a stock shape of material (a simple base feature) and add cut features and detail features, such as holes and rounds, just as a machinist would to the initial block of material.

The bottom line is that there are always several ways, and rarely one best way, to model a part. Think ahead about the easiest way to visualize the part, the most efficient method to extrude or revolve the cross section, the best way to add details, the critical dimensions of the part, what might be redesigned in the part, and how the part fits with other parts. Then proceed, understanding that there are many approaches that may work equally well.

8.2 MODELING THE HANDLE

The handle, shown in Figure 8.7, is a more complex part than the cap. The handle is generally cylindrical in shape with a circular hole at one end and a rectangular hole to accept the arms at the other end. Grooves on the handle provide better grip. One end is rounded and the other end has a reduced diameter. The handle could be made in several different ways. For instance, a circle could be extruded to form a cylindrical base feature, and then cuts could be added for the reduced diameter, rounded end, two holes, and grooves (a total of six features). Instead, you will model the handle by revolving a section that already includes the reduced diameter and rounded end. Then, you will add the grooves and holes (a total of four features).

Figure 8.7. Finished handle

8.2.1 Sketching the Base Feature of the Handle

You will begin by sketching the cross section of the handle to be revolved.

1. Open a new sketch for a new part and set it up as usual.
2. Activate the **Rectangle** tool by clicking the **Rectangle** button in the **Sketch Tools** toolbar, or **Tools** ⇒ **Sketch Entity** ⇒ **Rectangle**.
3. Bring the cursor over the origin and look for the square next to the cursor, indicating you are on the origin. Click and drag up and to the right. Release the mouse button. This creates a rectangle with its bottom left corner coincident with the origin.

4. Activate the **Line** tool. In the lower right corner of the rectangle, draw two lines, one horizontal and one vertical, to form a small rectangle (as shown in Figure 8.8). These two lines should start on the lines of the large rectangle and their endpoints should meet. Note that the symbol next to the pencil cursor changes when the cursor is over a line or a point.

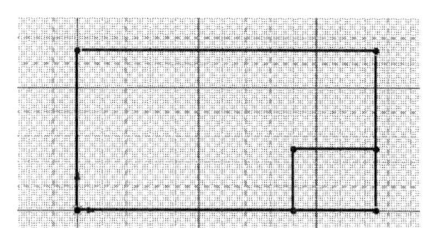

Figure 8.8.   Lines sketched

5. Activate the **Trim** tool by clicking the **Trim** button (a scissors) in the **Sketch Tools** toolbar, or **Tools** ⇒ **Sketch Tools** ⇒ **Trim**. The **Trim** tool clips away a part of a sketch entity. The bottom right corner of the rectangle will be clipped away.

6. With the **Trim** tool active, bring the cursor over the horizontal line segment on the right side of the rectangle, as indicated in Figure 8.9. The segment will be highlighted, indicating which part of the line will be trimmed away. Click on this segment to trim the line. Repeat the trim for the vertical line segment indicated in Figure 8.9.

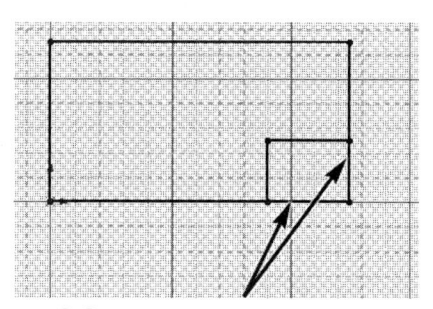

Figure 8.9.   Segments to be trimmed

7. Dimensioning a complex sketch can be challenging. In a sketch like this, you should place all of the dimensions first, and then change the values afterward. Activate the **Dimension** tool. Insert dimensions, as shown in Figure 8.10, by clicking on each line and placing the dimension box. The dimensions will probably be different from those in the Figure. Do not yet change the dimensions. After adding these four dimensions, the sketch is *Fully Defined*.

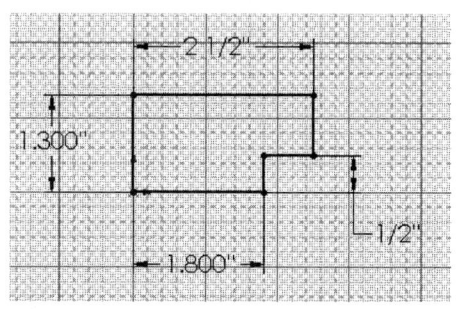

Figure 8.10.   Dimensions placed

8. Activate the **Select** tool, and modify each of the four dimensions to correspond to the values shown in Figure 8.11. As you change the dimensions, the aspect ratio of the sketch will change substantially. You might need to **Zoom In/Out** or **Zoom to Fit** to see the entire sketch.

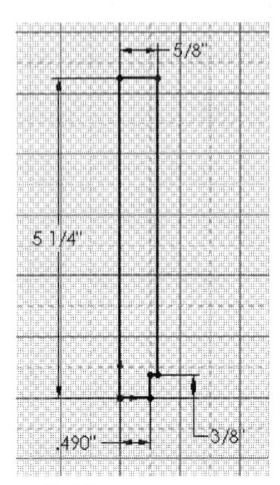

Figure 8.11. Dimensions updated

9. Now the round at the top right corner is added. Open the **Sketch Fillet** dialog box by clicking the **Fillet** button in the **Sketch Tools** toolbar, or **Tools ⇒ Sketch Tools ⇒ Fillet**. Enter *.3* for the **Radius**.

10. Move the cursor over the point at the upper right corner of the rectangle. Note that as the cursor is positioned over the corner, a circle appears next to the cursor. Click on the point. The corner is rounded and the .3 inch dimension is added. This is an alternate method available for adding a fillet. For previous fillets, the two line segments that intersected at the corner were selected instead of the corner point.

11. **Close** the dialog box and move the fillet dimension to a convenient position.

8.2.2 Revolving the Cross Section

Since this sketch will be revolved, you must to create an axis about which to revolve it.

1. Activate the **Centerline** tool.

2. Create a vertical centerline that starts at the origin and extends vertically upward. This completes the base sketch. Before continuing with the revolve, be sure that the screen looks similar to that shown in Figure 8.12.

3. Now that a centerline is in the sketch, the **Revolved Boss/Base** button becomes active in the **Features** toolbar. Click it in order to activate the *Revolve Feature* dialog box.

4. This is a closed sketch, so the revolve is a solid feature, not a thin feature. Only the *Revolve Feature* tab appears. Set the *Type* to *One-Direction* and the *Angle* to *360* degrees. This revolves the sketch 360 degrees in one direction from the plane of the sketch. Be sure that the *Revolve as* pull-down menu says *Solid Feature*.

Section 8.2 Modeling the Handle 133

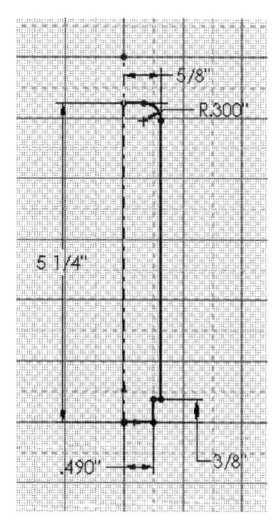

Figure 8.12. Fillet and centerline added

5. SolidWorks displays an outlined preview of the revolved section, but you may need to drag the **Revolve Feature** dialog box to a different location in order to see the preview. Click **OK** to accept the settings. Rotate the part and zoom in to see the part from all sides. In the isometric view, the part should look similar to that shown in Figure 8.13.

Figure 8.13. Handle base feature

8.2.3 Sketching and Cutting a Single Groove

The basic shape, or base feature, of the handle is complete, so details may now be added. The grooves in the handle are the first details to be added. Although the grooves appear to require many steps, they can be modeled easily by using a pattern feature. First, a single groove will be sketched and cut. Then, a pattern of grooves will be created based on this first groove.

1. Try to open a sketch by clicking on the **Sketch** button in the **Sketch** toolbar. SolidWorks will inform you that you need to select a planar face before sketching. Click **OK** to close the dialog box. The groove will be sketched on **Plane1**, so select it in the FeatureManager design tree. A green rectangle appears, showing **Plane1**. Now that you have selected a plane, open a sketch on it by clicking the **Sketch** button.

2. A new sketch opens and a grid appears. If the sketch plane is skewed in isometric orientation, double click **Normal To** in the **Orientation** dialog box. To see the sketch more clearly, click the **Hidden In Gray** button in the **View** toolbar.

3. In the first quadrant (above and to the right of the origin), draw a box using the **Rectangle** tool, as shown in Figure 8.14. This rectangle will become the groove, once you modify its shape, move it to the correct position and revolve it as a cut.

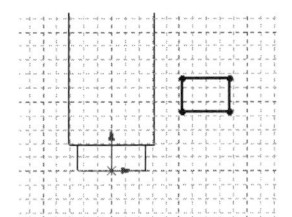

Figure 8.14. Rectangle drawn

4. First, position the groove vertically. With the **Dimension** tool, select the origin followed by the lower horizontal edge of the rectangle. Click again to place the dimension. With the **Select** tool, modify the dimension to be *1.5*.

5. Dimension both the height and width of the rectangle to be *.1*. Use **Zoom to Area** to zoom in on the rectangle, as shown in Figure 8.15.

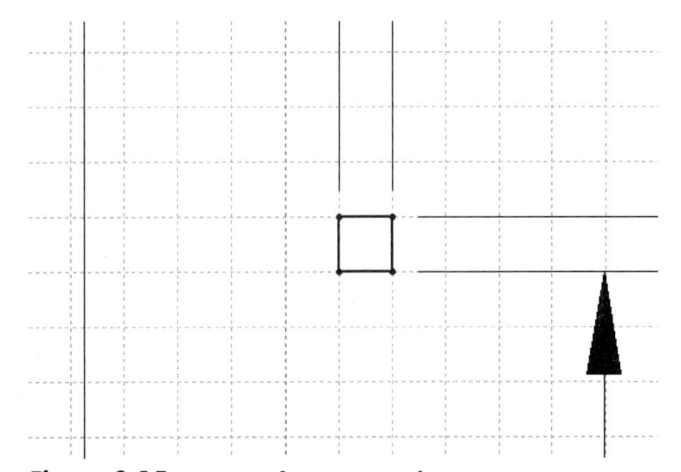

Figure 8.15. Zoomed in on rectangle

6. Open the ***Sketch Fillet*** dialog box by clicking the **Fillet** button in the **Sketch Tools** toolbar. Set the ***Radius*** to **.04**. Click the upper horizontal line of the rectangle, followed by the left vertical line. A fillet appears with a dimension. Before closing the dialog box, add another fillet to the lower-left corner of the rectangle by clicking on that corner point. The second fillet appears without a dimension, because the first dimension refers to both fillets. Close the ***Sketch Fillet*** dialog box. ***Zoom to fit*** and then move the fillet dimension to a convenient location with the **Select** tool.

7. Activate the **Centerline** tool. Draw a vertical centerline that starts at the origin. This centerline is the axis about which the cut for the groove will be revolved. You cannot use the centerline that was drawn for the previous sketch, since it is now part of the parent feature **Base-Revolve**, listed in the FeatureManager design tree. With the **Select** tool, click anywhere on a blank space in the sketch to ensure that nothing is currently selected.

8. The last thing to do is to place the rectangle where it belongs on the sketch with respect to the handle base feature. This could be done by making the dimension from the centerline to the right side of the rectangle the same as the radius of the handle. But if you cannot remember the radius of the handle, there is another method. You can use a relation to align the right side of the rectangle to the handle's outer edge. Use **Zoom To Area** to zoom in on the small rectangle. To add the relation, click on the **Add Relation** button in the **Sketch Relations** toolbar, or **Tools** ⇒ **Relations** ⇒ **Add**. The ***Add Geometric Relations*** dialog box appears. Then, click on the right side of the rectangle. **Zoom to Fit** and then click on the right edge of the handle. Both lines appear in the ***Selected Entities*** field of the dialog box and both are highlighted. The edge of the handle is called ***Silhouette Edge <1>***, because the selection is really the silhouette of a curved surface of the handle. The selected items should be collinear, so be sure that the **Collinear** button is selected. Then, click ***Apply*** to make the two lines collinear. With this relation, the groove will always be collinear with the edge of the handle, even if the handle's radius is changed. The sketch should be similar to Figure 8.16. **Close** the ***Add Geometric Relations*** dialog box.

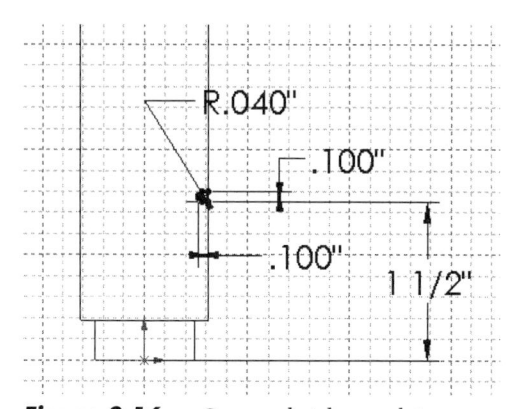

Figure 8.16. Groove sketch complete

9. The groove can be made by revolving the small rectangle as a cut to remove material around the handle. Click the **Revolved Cut** button in the **Features** toolbar, or **Insert** ⇒ **Cut** ⇒ **Revolve**. Since material is being removed from the part, a **Revolved Cut** is used, not a **Revolved Boss/Base**, which creates material.
10. The *Revolve Feature* dialog box opens. This is the same dialog box that appears for a **Revolved Boss/Base**, except that material is removed with a **Revolved Cut**.
11. Revolve the sketch all the way around the centerline. Click **OK** to accept the settings. In the isometric view, the part should look similar to that shown in Figure 8.17.

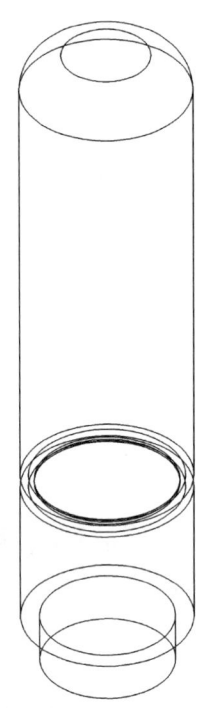

Figure 8.17. Groove cut

8.2.4 Modeling More Grooves: Inserting a Linear Pattern

Instead of re-creating the previous sketch again and again to make more grooves, you can use a linear pattern to repeat the revolved cut several times along the length of the handle to create a pattern of grooves.

1. The pattern of grooves must be repeated along a defined direction. To do this, an axis will be created from the intersection of two planes. Click on **Plane1** in the FeatureManager design tree and then *control click* **Plane3** to select both planes. The intersection of **Plane1** and **Plane3** is a vertical line.
2. Click **Insert** ⇒ **Reference Geometry** ⇒ **Axis** to create an axis at the intersection of the two planes. The *Reference Axis* dialog box appears, as shown in Figure 8.18.
3. The *Two Planes* button should be already be selected. *Plane1* and *Plane3* should appear in the *Selected Items* list. Click **OK** to create the axis. Verify that **Axis1** is the axis of the handle by rotating the handle.

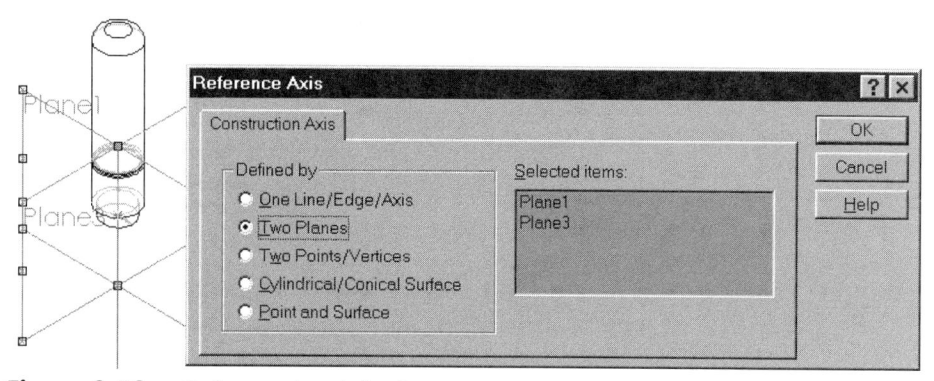

Figure 8.18. Reference Axis dialog box

4. A pattern will be made from the revolved cut that created the groove. Click on **Cut-Revolve1** in the FeatureManager design tree to select it. The cut becomes highlighted in green.

5. Click the **<u>Linear Pattern</u>** button (a 2 × 3 array of boxes) in the **Features** toolbar, or **Insert** ⇒ **Pattern/Mirror** ⇒ **Linear Pattern**, to create a pattern based on the highlighted feature. The **Linear Pattern** dialog box appears, as shown in Figure 8.19.

6. **Cut-Revolve1** should appear in the **Items to copy** list. If it does not, select it in the FeatureManager design tree.

7. The **Direction Selected** box should be highlighted in pink. If it is not, click in this box so that the next selection will indicate the direction. Click **Axis1** in the FeatureManager design tree to specify the direction in which the pattern will be created. The axis should be highlighted in green. An arrow appears on the handle, showing the direction of the pattern creation. The arrow should point upward, toward the rounded end of the handle. If the arrow is pointing downward to the squared-off end, check the **Reverse Direction** checkbox.

8. Set the **Spacing** to **.25**. This is the distance between each instance of the pattern.

9. Click the up arrow in the **Total Instances** field until it shows **8**, for a total of eight grooves. A preview is shown as the number of cuts is increased. Be sure that the screen looks similar to that shown in Figure 8.19.

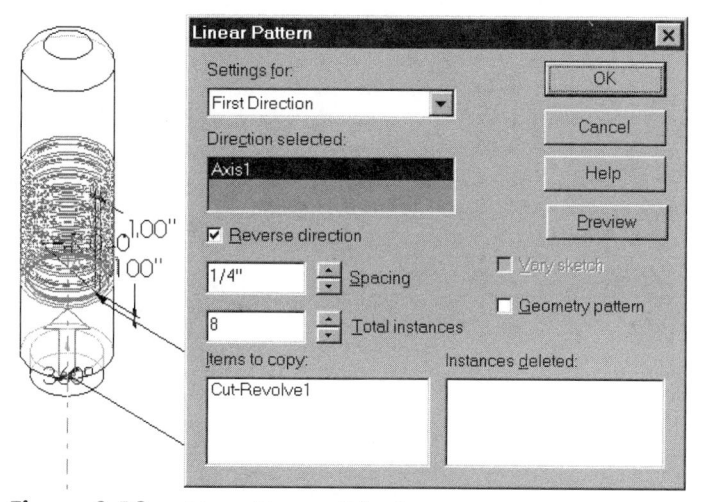

Figure 8.19. Linear Pattern dialog box

10. Click **OK** to accept the settings. Eight grooves are created along the length of the handle.
11. **Axis1** will no longer be needed, so *right click* it in the FeatureManager design tree. Select **Hide** from the menu so that **Axis1** is not shown. Shade the part and rotate it in order to see your progress. The handle should look like Figure 8.20, in the **Front** orientation using the **Shaded** view.

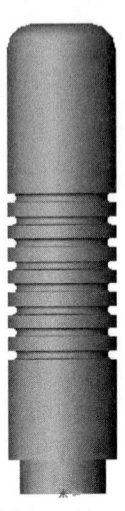

Figure 8.20. Grooves patterned

8.2.5 Modeling the Rectangular Hole

A rectangular hole at the bottom end of the handle holds the two arms. An extruded cut will be used to make this hole.

1. Open a sketch on the bottom surface of the handle by rotating it in the **Hidden In Gray** display so its flat end is visible. Select the bottom face (not the edge), as indicated in Figure 8.21. Click the **Sketch** button in the **Sketch** toolbar to open a sketch in the plane of the bottom face of the handle.

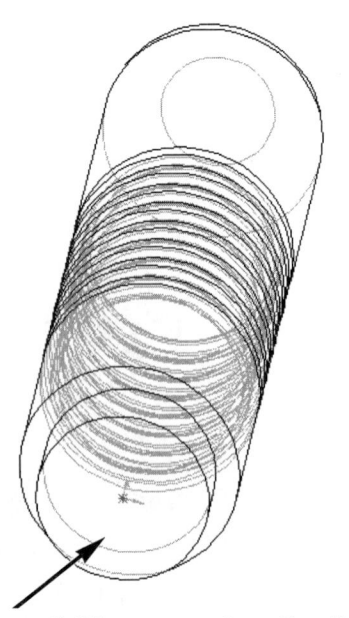

Figure 8.21. Bottom face of handle

Section 8.2 Modeling the Handle **139**

2. Reorient the sketch plane by double clicking **Normal To** in the **Orientation** dialog box.
3. Draw a rectangle that surrounds the origin. Be careful: If any of the vertices lie on the origin or on the arc of the circle, you may have created an undesirable constraint that will cause problems when dimensioning later. If necessary, delete the rectangle and draw a new one.
4. Dimension both the geometry and location of the rectangle, as shown in Figure 8.22. It may be necessary to **Zoom To Fit**.

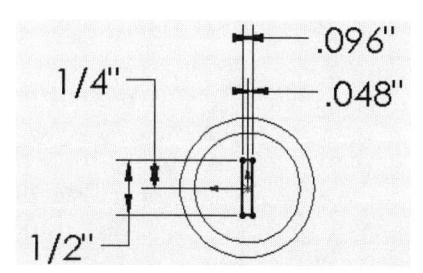

Figure 8.22. Rectangle hole sketch complete

5. Some of the dimensions are difficult to read because the arrowheads overlap. To correct this, select the numerical portion of the dimension. Then click one of the **Arrow In\Out\Smart** buttons in the PropertyManager that is now visible where the FeatureManager design tree is normally located.
6. Cut a blind hole that extends **3/4** into the handle. Before clicking **OK** in the **Extruded Cut Feature** dialog box, rotate the part and look at the preview, to be sure that the hole is going into the handle. The rectangular end of the hole should be apparent. If it looks as if you are cutting in the wrong direction, click the **Reverse Direction** checkbox. After clicking **OK**, the part should look like Figure 8.23.

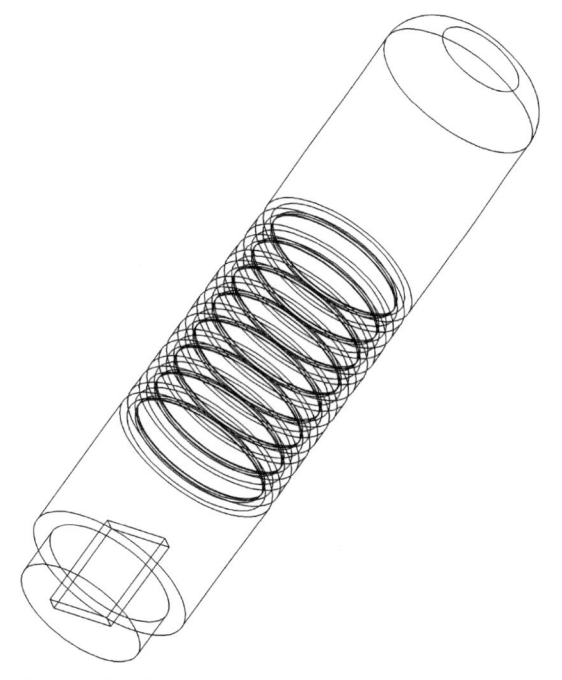

Figure 8.23. Rectangular hole cut

8.2.6 Modeling the Circular Hole

The next feature to add is the circular hole at the rounded, top end of the handle. It is perpendicular to the widest dimension of the rectangular hole at the opposite end of the handle. The trick here is to determine which plane should be used as the sketch plane.

1. View the handle, using the **Hidden In Gray** display. Click on each plane in the FeatureManager design tree to see which plane is parallel to the long side of the rectangular cut. It may help to rotate the handle to look toward the flat. You should find that **Plane1** is parallel to the long side of the rectangular cut. Click **Plane1** in the FeatureManager design tree and open a sketch on it. Orient the sketch so that you are normal to the sketching plane by double clicking **Front** or **Normal To** in the **Orientation** dialog box.

2. Centerlines are often useful in sketching. In this case, draw a vertical centerline extending from the origin to the top of the handle. The lightbulb next to the cursor indicates that the centerline is coincident with a previous centerline.

3. Draw a circle near the top of the handle with its center point on the centerline, as shown in Figure 8.24. This ensures that the circle is directly above the origin.

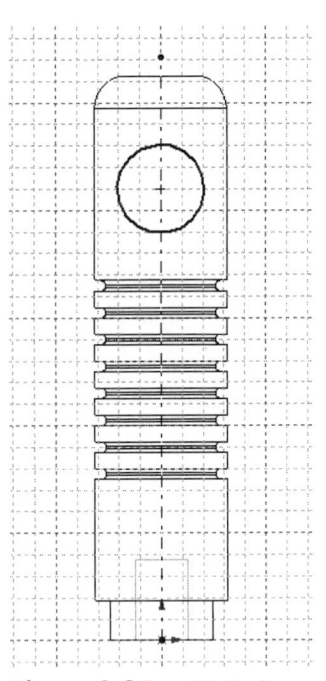

Figure 8.24. Circle drawn

4. Dimension the diameter of the circle to be **1/2**, and the distance between the center of the circle and the origin to be **4 1/2**.

5. Open the **Extrude Cut Feature** dialog box. The hole should go through the handle on both sides of the sketch plane. The sketch plane lies at the center of the handle, so check the **Both Directions** check box. Currently, the **Settings for** field indicates **Direction 1**. Set the **Type** to **Through All**, so that the cut goes through all the material of the handle from the sketch plane outward. Rotate the handle using the **Rotate View** tool. Notice that a black circle appears at the center of the handle (where the hole will be) and a yel-

low circle appears on one side of the handle. The yellow circle indicates the side of the sketch plane to which the **Through All** setting applies. Now, change the **Settings for** field to **Direction 2**, and notice that the yellow circle flips to the opposite side of the handle. Set the **Type** to **Through All** for the cut on the opposite side of the sketch plane. Note that different types of cuts can be applied on each side of the sketch plane when the cut is in **Both Directions** from the sketch plane. In this case, the material will be removed **Through All** on both sides of **Plane1**.

6. Click **OK** to accept the settings. The result should look like Figure 8.25. Rotate the part to be sure that the hole does indeed go all the way through the handle.

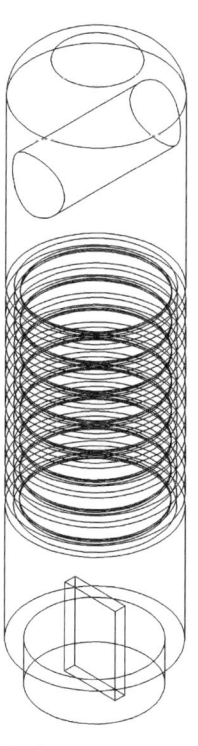

Figure 8.25. Circular hole cut

8.2.7 Adding the Finishing Touches to the Handle

Creating four rounds, or fillets, is all that is needed to complete the model of the handle. A round will be added on each side of the circular hole and on each of the edges at the flat bottom of the handle.

1. The rounds on both ends of the circular hole are identical. Consequently, they will be added as a single feature. If the radius is changed, only one feature needs to be updated. In the **Hidden In Gray** display, rotate the handle, so that one end of the circular hole is visible. Since you are filleting a feature and not sketch entities, use the **Fillet** button in the **Features** toolbar. Type in the radius of **.050**, and select the edge of the circular hole that is visible on one side of the handle. Then, click the edge of the circular hole that is shown as a hidden line on the other side of the handle. Both **Edge <1>** and **Edge <2>** should appear in the **Edge Fillet Items** box. Click **OK** to finish. The result should look like Figure 8.26.

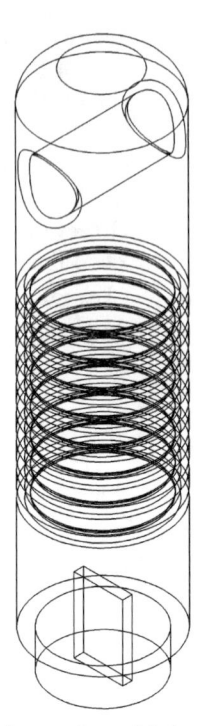

142 Chapter 8 Modeling Parts in SolidWorks: Revolves

Figure 8.26. Fillets added to circular hole

2. Now, the edges at the flat bottom of the handle will be rounded. Rotate the view so that the bottom flat end is visible. Add a **.075** fillet to one of the edges at the bottom end of the handle, as indicated in Figure 8.27. Do not control-click to round both edges at the same time. This permits the rounds to be altered separately later on, if necessary.

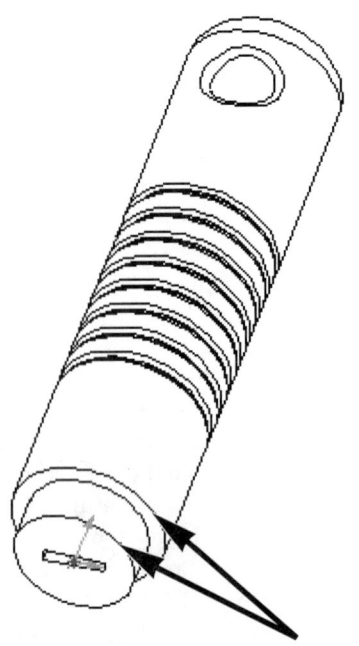

Figure 8.27. Edges to be rounded

3. Add a **.075** fillet to the other edge, as indicated in Figure 8.27. The final results should look like Figure 8.28 in **Hidden Lines Removed** display.

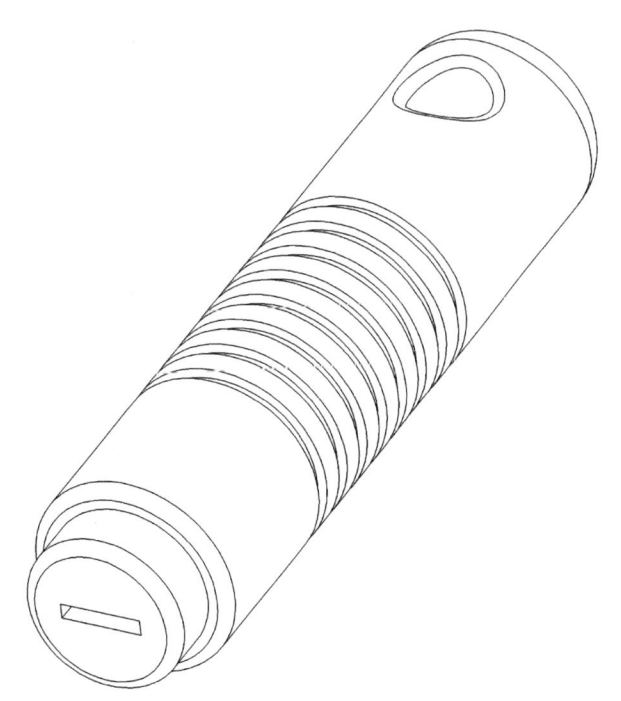

Figure 8.28. Rounds added

8.2.8 Changing the Color of the Handle

The color of parts can be changed to reflect the part's material or to make the part appear more realistic. Set the view to **Shaded** to see the color of the handle. The default color is a metal-like gray color. This color is acceptable for metal parts such as the guard, arm, blade, and cap. The color of the handle will be changed since it would most likely be made of wood or plastic.

1. Select **Tools ⇒ Options**. Click on *Colors*.

2. The standard system colors can be changed using the initial dialog box that appears. However, you need to change the color of the part so click the **Go To Document Colors** button. Click *Shading* to highlight it, as shown in Figure 8.29. Then click the **Edit** button.

3. Choose a color for the handle from the swatches that are shown in Figure 8.30 and then click **OK**.

4. Click **OK**. The color of the handle has changed to the new color. Experiment with colors to find a color that looks realistic for the handle. Be sure that the color is not too dark, so the details of the part not will be difficult to see.

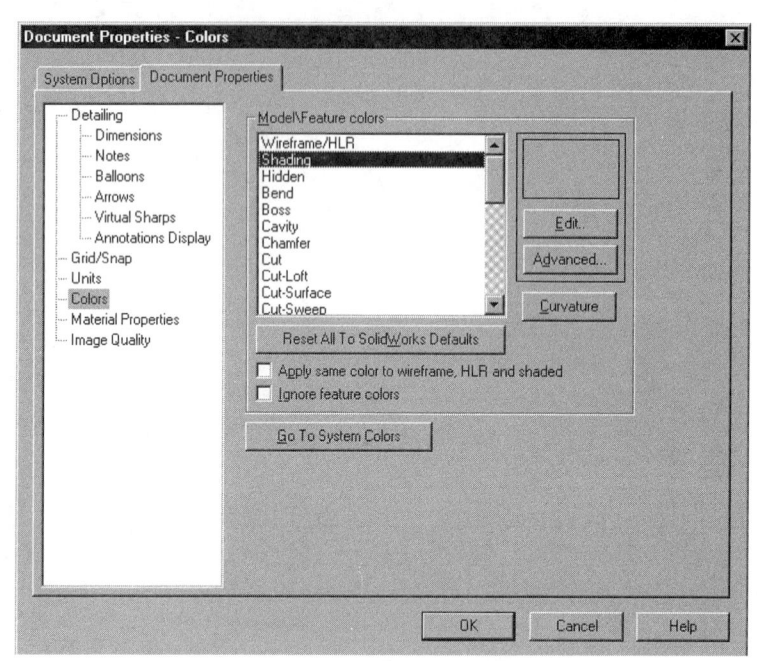

144 Chapter 8 Modeling Parts in SolidWorks: Revolves

Figure 8.29. Document properties dialog box: color

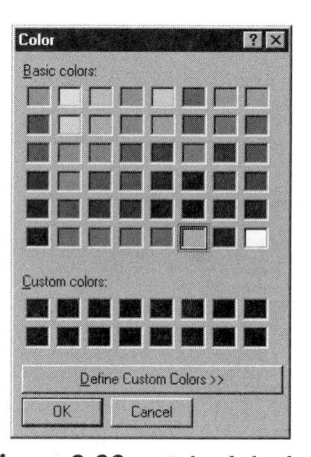

Figure 8.30. Color dialog box

Congratulations! You have completed modeling the handle, a complicated part. Save the part as ***handle***. Rotate the handle to see that it looks like the one shown at the beginning of this section. Close the window.

8.3 MODELING THE RIVET

The last part of the pizza cutter to model is the rivet that secures the blade between the two arms. The rivet will be modeled to look as it would after being deformed to fasten the cutting blade to the arms. The rivet may look complicated in the three-dimensional views shown in Figure 8.31. However, its shape is more evident when shown in cross section as in Figure 8.32. In the planning stages, imagining the cross-section of a part helps in modeling complex shapes.

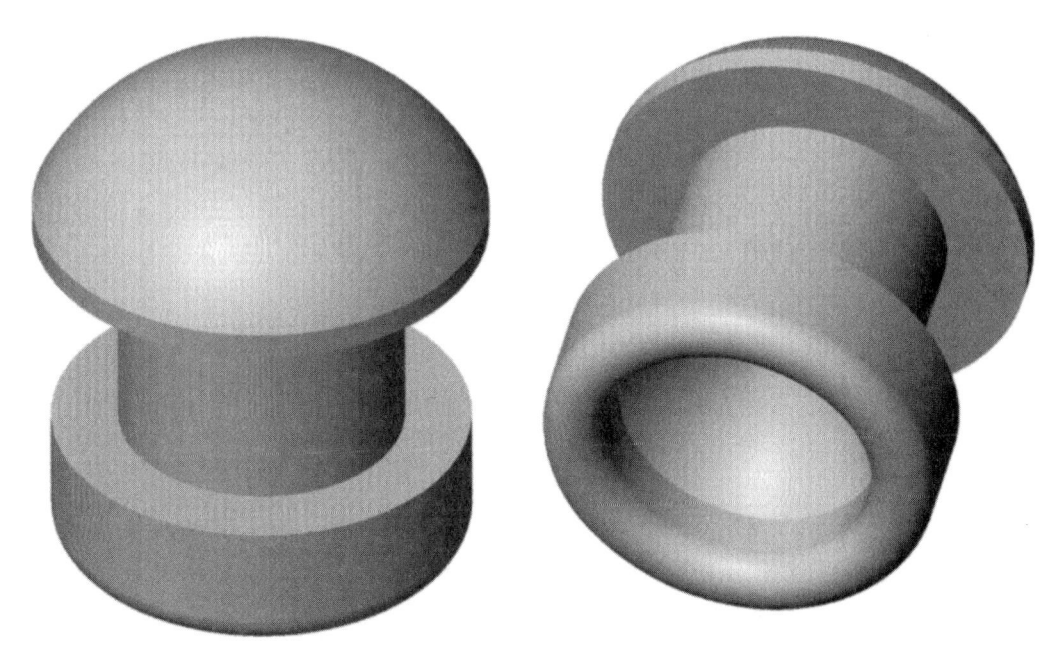

Figure 8.31. Finished rivet

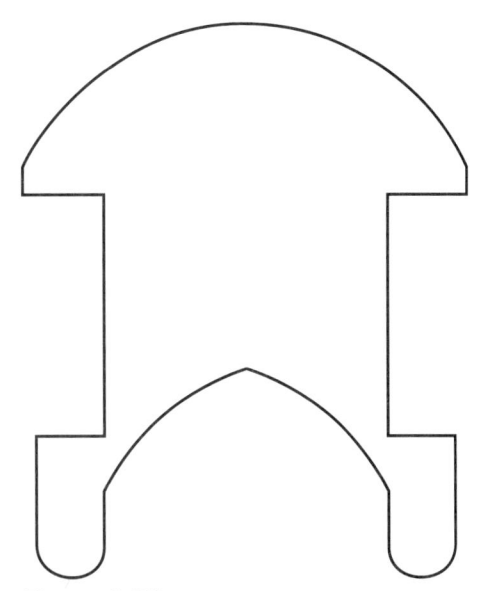

Figure 8.32. Rivet cross section

The rivet is symmetric about a vertical axis. Taking advantage of this symmetry makes it easier to model the part. The section shown in Figure 8.33 can be revolved about a vertical axis to model the rivet. You will add the round at the bottom of the rivet to complete the part using only two features (a revolved section and a fillet). This part could be modeled in a single feature, by adding the bottom round before the section is revolved. However, the section sketch to be revolved is already complicated, so the fillet will be added after revolving the section.

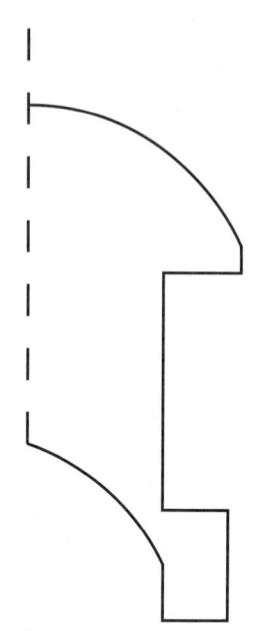

Figure 8.33. Rivet sketch, undimensioned

8.3.1 Creating Arcs

The cross section of the rivet, shown in Figure 8.33, is based on arcs. There are three types of arc tools available in the **Sketch Tools** toolbar. Each arc tool uses a different geometric method to create an arc.

- A **Centerpoint Arc** starts with a point at the center of the arc. Then, both a radius and an angle are defined by selecting the start and end points of the arc.
- A **Tangent Arc** is tangent to an existing sketch entity.
- A **3 Point Arc** is defined by three points that lie on the arc (start point, end point, and midpoint).

You will be using a **3 Point Arc** and a **Centerpoint Arc** to sketch the cross section in Figure 8.33 that will be revolved to model the rivet. The first arc is the one at the bottom of the rivet.

1. Open a new sketch for a new part and set it up as usual.
2. Click **3 Pt Arc** in the **Sketch Tools** toolbar, or **Tools** ⇒ **Sketch Entity** ⇒ **3 Point Arc**. A semicircle appears next to the cursor to indicate that an arc will be sketched.
3. Start the left end of the arc by clicking on the origin and dragging the cursor down and to the right. Release the mouse button. This defines the location of the endpoints of the arc, but the arc is not yet finished. A third point must be defined. The screen should look like that shown in Figure 8.34, with a dashed arc and a green point at the midpoint of the arc.
4. Click and drag the green point. As you drag the point, notice that both the angle and radius of the arc are shown next to the cursor. Release the mouse button to finish the arc so that it is similar in shape to the arc in Figure 8.34. Be sure that the center of the arc (designated by a small cross) is not directly below the origin. If it is, delete the arc and redraw it.

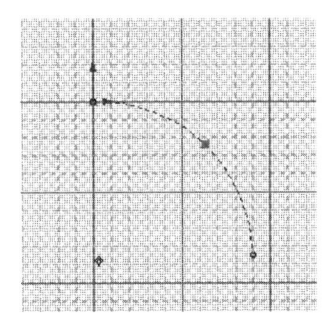

Figure 8.34. Arc drawn

8.3.2 Drawing the Rest of the Sketch

1. Activate the **Line** tool, and draw the seven line segments, as shown in Figure 8.35, to outline the shape of the rivet, starting at the bottom of the arc. Be sure that each line is either horizontal or vertical, and that the endpoints of the lines are coincident.

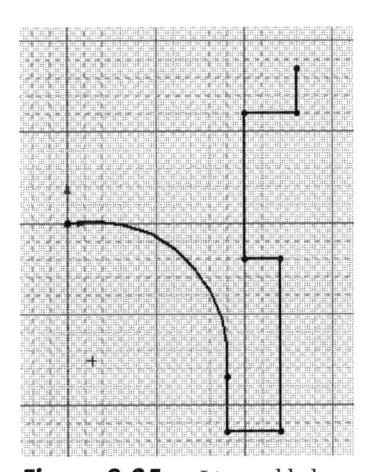

Figure 8.35. Lines added

2. If necessary, use the **Zoom In/Out** tool, so that there is space above the upper line segment that was just drawn. Then, draw a vertical **Centerline** that starts at the origin and extends vertically upward, well past the upper end of the line segments that were just sketched. This is the centerline for revolving the sketch.

3. The top arc of the sketch starts at the upper endpoint of the upper vertical line. The arc extends to the centerline and has a center point (denoted by the small cross) that lies on the centerline, as shown in Figure 8.36. Click the **Centerpoint Arc** button in the **Sketch Tools** toolbar, or **Tools ⇒ Sketch Entity ⇒ Centerpoint Arc**. Bring the cursor over the centerline. A lightbulb appears next to the cursor to indicate that it is pointing to the centerline. Click and hold the left mouse button at a point that is both on the centerline and lies just above the origin. This places the center point.

4. Drag the cursor to the upper endpoint of the upper vertical line. Look for the square indicating that the cursor is above the point. Release the mouse button. This defines the first endpoint of the arc. Use **Zoom to Fit** if necessary in order to see the entire arc.

5. To define the second endpoint of the arc, click and hold the left mouse button anywhere on the sketch. Drag the cursor to the centerline and release the

mouse button. This places the second endpoint on the centerline, directly above the origin. Be sure that the arc has one endpoint on the centerline and one endpoint on the upper end of the vertical line. Also, be sure that the arc's center point lies on the centerline. The sketch should look similar to Figure 8.36.

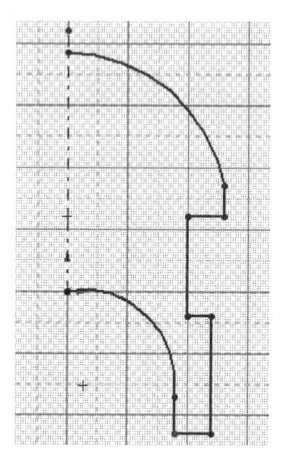

Figure 8.36. Second arc added

6. Create a closed sketch by drawing a vertical line that extends from the endpoint of the first arc to the endpoint of the second arc along the centerline. Since the section is a closed shape, it can be revolved to model a solid object. Note that the sketch is *Under Defined*, since no dimensions have yet been added. You will add dimensions after revolving the sketch.

8.3.3 Revolving the Sketch

The rivet is modeled by revolving the sketch around the centerline. For previous parts, you were careful to constrain the sketch, using dimensions and relations, before revolving the section. In this case, the section will be revolved first, and the dimensions to fully define the model will be added later.

1. To revolve the sketch, click the **Revolved Boss/Base** button in the **Features** toolbar.
2. Revolve the sketch 360 degrees around the vertical centerline. The isometric view of the part should look similar to that shown in Figure 8.37.

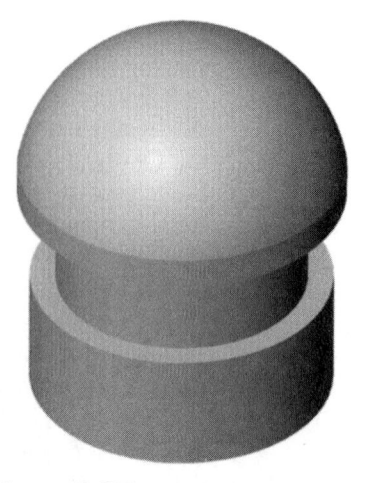

Figure 8.37. Undimensioned rivet

8.3.4 Reopening the Sketch and Adding Dimensions

The revolve created a three-dimensional object without fully defining the sketch. You can go back and add dimensions to the sketch in order to constrain the geometry. The ability to make three-dimensional objects without fully defining a sketch is a useful tool for quickly displaying the approximate geometry of a part.

1. In the FeatureManager design tree, click on the plus sign to the left of **Base-Revolve**. This displays the components of the revolve feature, namely **Sketch1**.
2. *Right click* **Sketch1**, and the sketched cross section that was revolved will be highlighted. Choose **Edit Sketch** from the menu. This hides the revolved feature and opens the sketch that was created before revolving the cross section.
3. *Double click* **Normal To** in the **Orientation** dialog box, so that the sketch plane is the plane of the screen.
4. With the **Dimension** tool, add dimensions to the radii of the arcs. This is done by clicking on each arc and then clicking to place the dimension. The radii should appear as in Figure 8.38. The "R" in front of the dimension indicates that the dimension describes the radius of the arc, not its diameter.
5. With the **Dimension** tool still active, place the six vertical dimensions and the three horizontal dimensions, as shown in Figure 8.38. It may help to **Zoom In/Out** as you add dimensions. The values shown in the Figure will probably be different than the values in your sketch. When dimensioning, remember that you can dimension from point to point, line to line, or just dimension the length of the line. After placing the dimensions, the sketch should be fully defined.

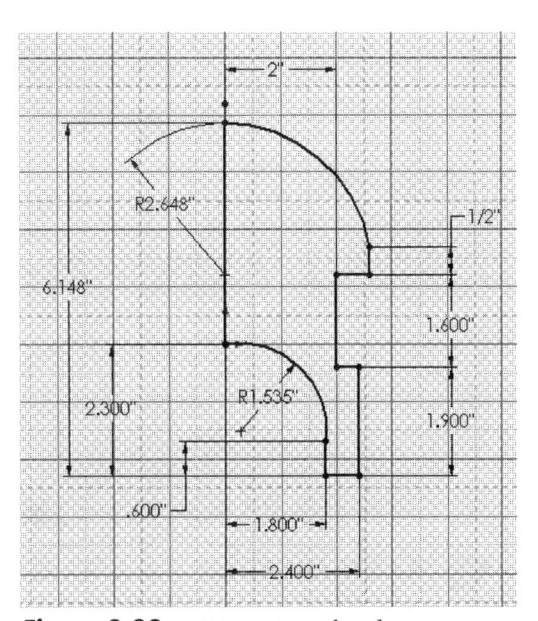

Figure 8.38. Dimensions placed

6. The dimensions that are now on the sketch are much larger than the ones necessary for the final part. For example, the height of the rivet is only .32 inches, but the present value is much larger (6.148 inches for the example in Figure 8.38). Changing all of the dimensions at once, rather than one at a time, keeps the sketch in proportion as the dimensions are modified. Uncheck the

Automatic Solve feature in **Tools** ⇒ **Sketch Tools**. When this feature is unchecked, SolidWorks will not immediately change the length of the line when you modify the dimension value. Notice that the Status bar displays **Auto-Solve Mode Off**.

7. Change each dimension as you normally would in order to match the values to those in Figure 8.39. With automatic solve off, the geometry of the part does not change to correspond to the new dimensions.

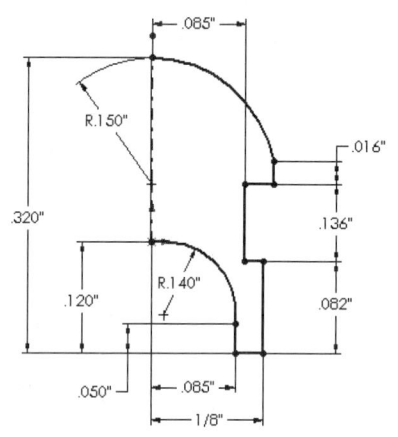

Figure 8.39. Dimensions updated

8. Be sure that all of the dimensions have been changed. Now that all of the dimensions reflect the desired values, click **Tools** ⇒ **Sketch Tools** ⇒ **Automatic Solve**, so that **Automatic Solve** is checked. The sketch adjusts itself to the smaller values. Zoom in to be sure that the part's geometry is similar to that shown in Figure 8.40. If it is not, undo the last operation and be sure that all of the dimensions are correct. You may need to move some of the dimensions and **Redraw** the sketch. Do not worry about the large size of the dimensions and arrows for now. Just be sure that the shape of the cross section is correct.

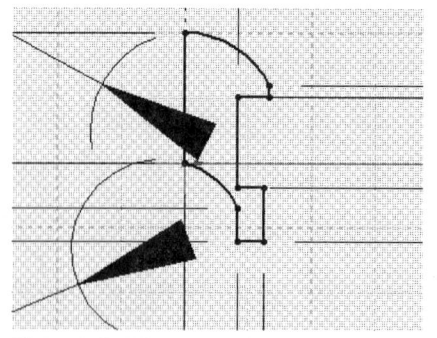

Figure 8.40. Sketch solved

8.3.5 Changing the Detailing Settings

The font size of the dimensions and the size of the arrows are much too large for the sketch as it is. To make it easy to see each of the dimensions, you can reduce their sizes, using the **Detailing** option tab of the **Document Properties** dialog box.

1. Click the **Grid** button to open the **Document Properties** dialog box. Select the **Detailing** option in the menu on the left side of the dialog box. The values in the **Detailing** option will be set to match those shown in Figure 8.41.

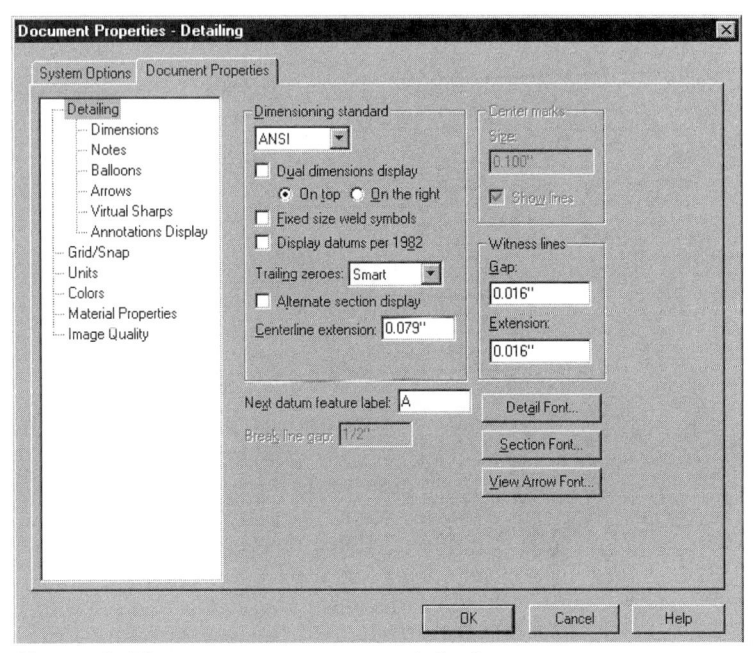

Figure 8.41. Document Properties dialog box

2. In the **Witness lines** section,

    - set the **Gap** between the end of the extension line and the visible line to *1/64*, and

    - set the **Extension** of the witness lines past the dimension to *1/64*.

3. Click on the **Dimensions** option in the menu on the left side of the dialog box.

    - Click the **Font** button. Set the **Units** to *.025* for the height of the font and click **OK**.

    - Set the **Bent leader length** to *.050*.

4. Click on the **Arrows** option in the menu on the left side of the dialog box. In the **Size** section, set

    - the short dimension of the arrowhead by setting the **Height** to *.007*,

    - the long dimension of the arrowhead by setting the **Width** to *.020*, and

    - the **Length** of the line of the dimension to *.050*.

5. Click **OK** to accept the settings. The dimensions are now proportional to the sketch. After arranging the dimensions, the sketch should look similar to that shown in Figure 8.42. You may need to zoom in or out, or move the dimensions several times to get the sketch to look right. The centerline can be shortened by dragging the endpoint with the **Select** tool.

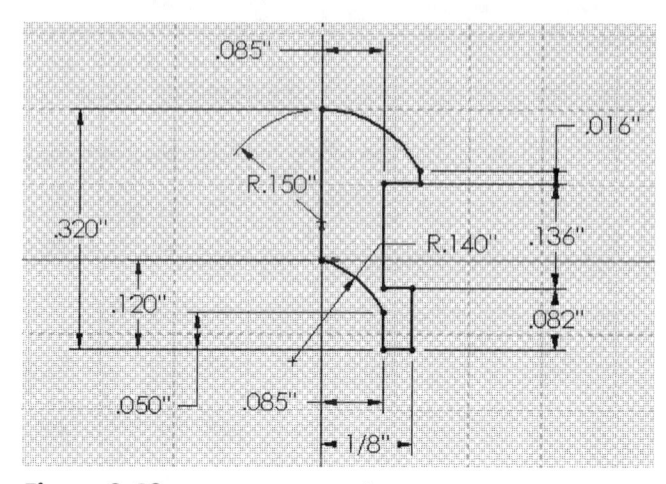

152 Chapter 8 Modeling Parts in SolidWorks: Revolves

Figure 8.42. Dimensions sized

6. To exit the sketch, click on the **Sketch** button in the **Sketch** toolbar to turn off the sketch mode. The rivet is automatically rebuilt using the new sketch dimensions.

8.3.6 Rounding the Bottom of the Rivet

1. Previously, you had added fillets to round edges by selecting the edge. Fillets can also be used to round a face of a part. When a face is filleted, all of the edges that bound the face are rounded. To round the bottom of the rivet, rotate the part so that the bottom of the rivet can be seen.
2. Use the **Select** tool to highlight the flat face that forms the bottom of the rivet. The face becomes highlighted.
3. Click the **Fillet** button in the **Features** toolbar. *Face <1>* should appear in the *Fillet Feature* dialog box.
4. Set the **Radius** to **.02** and click **OK**. The bottom of the rivet should look like Figure 8.43.

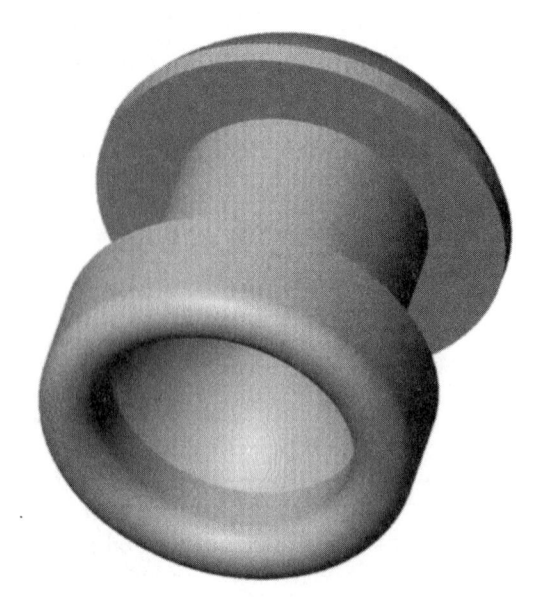

Figure 8.43. Finished rivet, underside

This completes the rivet. Save the part as ***rivet*** and close the window.

Congratulations! You have successfully modeled all six parts of the pizza cutter. The next chapter will step you through the assembly mode of SolidWorks to assemble these parts into a pizza cutter. The following chapter will describe how to create two-dimensional drawings of both the parts and the assembly.

PROFESSIONAL SUCCESS: SWEEPS AND LOFTS

SolidWorks allows the creation of smooth, curved features using sweeps and lofts. Sweeps are used to create features that have a cross section defined along a trajectory. An example of a sweep is the handle of a ceramic coffee mug. The cross section of the handle is usually oval in shape. A mug handle can be created by "sweeping" the oval cross section through a U-shaped trajectory.

Before a sweep is created, the cross section and the trajectory are sketched as two different sketches. A sweep is implemented by using the **Sweep** button in the **Features** toolbar, or **Insert** ⇒ **Base** ⇒ **Sweep**. The cross section sketch is set as the ***Sweep section***, and the trajectory sketch is set as the ***Sweep path***. The cross section can then be swept along the defined trajectory.

Lofts are smooth transitional surfaces between at least two planar sections. For instance, imagine a ventilating duct that transitions from a circular cross section to a rectangular one. A loft is used to create the transition surface of the duct. A loft feature is created by sketching two cross sections, each in a different sketch plane, and then clicking the **Loft** button in the **Features** toolbar, or **Insert** ⇒ **Base** ⇒ **Loft**.

Both sweeps and lofts can be used to remove material instead of create material. These features are created in the same way, except that they are found in the **Insert** ⇒ **Cut** menu.

KEY TERMS

Arc
Centerline
Construction geometry
Pattern
Revolve
Trim

Modeling Problems

1. Model the blade using a revolve instead of an extrusion. Using the proper sketch, the entire blade can be modeled using a single revolve and no other features.
2. Model the cap as a thin revolved section without the hole. Add the hole as a circular, extruded cut.
3. Measure the dimensions of a nail, and then model it in two different ways:

    a. Revolve a single cross section.
    b. Extrude a cylinder size of the head, and then revolve cuts to make the shaft of the nail and sharp point.

4. Model the handle by revolving a sketch of a rectangle. After revolving the section, revolve a cut at the bottom end of the handle to model the reduced diameter. Then, round the top end of the handle. Finish the handle by adding the holes and grooves.
5. Model the handle by extruding a circle along its axis to create the base feature. Then, model the reduced diameter at the rectangular-hole end of the handle using a revolved cut. Add the grooves as a revolved feature, and finish by adding the holes and the rounds.
6. Model a hollow handle with an outer geometry identical to the handle in this chapter. The hollow core should have a uniform diameter of .5 inches. Omit the rectangular hole at the bottom of the handle

9

Modeling an Assembly: The Pizza Cutter

OVERVIEW

A key feature of SolidWorks is the capability to assemble parts in a virtual environment. This feature allows an engineer to check fits and interference between parts and to visualize the overall assembly. Assembly involves orienting parts with respect to each other. Once assembled, the interference between parts can be checked. Then, errors in the design of individual parts can be corrected.

9.1 MODELING THE CUTTER SUB-ASSEMBLY

One of the great advantages of SolidWorks is that it allows engineers or designers to assemble modeled parts in order to create a virtual assembly. This helps the designer to work out problems with part interactions early in the design phase, before manufacturing has begun. SolidWorks has features that automatically check for both clearances and interference between parts in an assembly. It also permits the engineer or designer to get an idea of what the assembled object looks like before it is fabricated.

In SolidWorks, an assembly begins by bringing a single part into an assembly document and orienting it with respect to the planes of the assembly. Then, additional parts are brought into the document one at a time and oriented with respect to the first part. Once all of the parts have been assembled, the assembly can be checked for interference, or overlap, between parts.

SECTIONS

- 9.1 Modeling the Cutter Sub-Assembly
- 9.2 Modeling the Pizza Cutter Assembly
- 9.3 Checking the Assembly for Interference
- 9.4 Creating an Exploded View of the Assembly

OBJECTIVES

After working through this chapter, you should be able to

- Open a new assembly
- Create a sub-assembly
- Bring parts into an assembly or sub-assembly
- Orient and constrain parts with respect to datum planes and other parts
- Check for interference in an assembly
- Correct errors in parts while assembling
- Create an exploded view of an assembly

In this chapter, you will use the parts modeled in the previous sections to make an assembly of the pizza cutter. Do not begin this section until all parts of the pizza cutter have been modeled. First, you will model the cutter sub-assembly, which consists of the blade, the rivet, and two arms. The final cutter sub-assembly is shown in Figure 9.1.

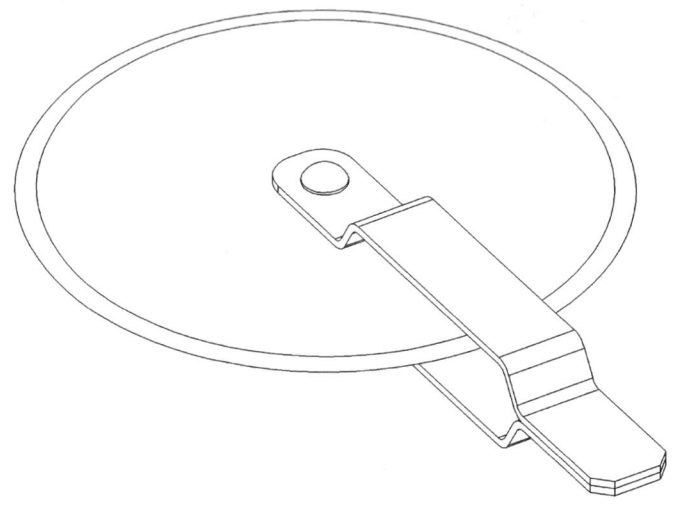

Figure 9.1. Finished cutter sub-assembly

9.1.1 Creating a New Assembly Document

The assembly process begins with a new document, called an assembly document, which differs from a part document.

1. With the SolidWorks window open, click **File ⇒ New**.
2. Instead of choosing to make a new part at this point, select **Assembly** in the dialog box and click **OK**.
3. A new assembly window opens, named **Assem1**. Note the seven items present in the FeatureManager design tree: annotations, lighting, three datum planes, the origin, and MateGroup1 (related to mating parts together).
4. In the **View ⇒ Toolbars** menu, be sure that the **Standard**, **View**, **Assembly**, **Features**, **Selection Filter**, **Sketch**, **Sketch Relations**, and **Sketch Tools** are checked, so that these toolbars appear on the screen.
5. The **Selection Filter** toolbar, shown in Figure 9.2, is used to restrict the type of items that can be selected. If the entire toolbar is not visible, click on the double bars at the left end of the toolbar, drag the toolbar to just below the **View** toolbar, and release the mouse button. Each toolbar button represents a particular type of entity. When a part is clicked with the **Select** tool, only the types of items activated in the **Selection Filter** toolbar can be selected. This is helpful in selecting the desired item during assembly. For instance, if the **Filter Faces** button is active, the **Select** tool will only select faces (surfaces) of the part. Move the cursor over each button in the **Selection Filter** toolbar to see what items can be selected. One or several but-

tons can be activated at a time. Since faces are most commonly used to orient parts in an assembly, activate the **Filter Faces** button to allow only faces to be selected.

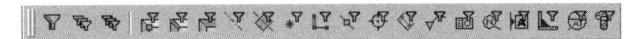

Figure 9.2. Selection filter toolbar

6. In the **View** toolbar, click the **Hidden Lines Removed** button. This makes it easier to decipher the various components of the assembly.

9.1.2 Bringing the Rivet into the Sub-assembly

Assemblies and sub-assemblies can consist of many parts. The first part inserted into the assembly is usually the foundation for all of the parts that are added later. In this case, you will use the rivet as the first part.

1. To insert the rivet into the assembly, select **Insert ⇒ Component ⇒ From File**.

2. Go through the file structure and find the rivet, as indicated in the dialog box that is shown in Figure 9.3. Be sure the *Files of type* pull-down menu is set to *Part Files*, so that the rivet file is displayed. Highlight the file named **rivet** (which was modeled in the previous chapter) and click **Open**.

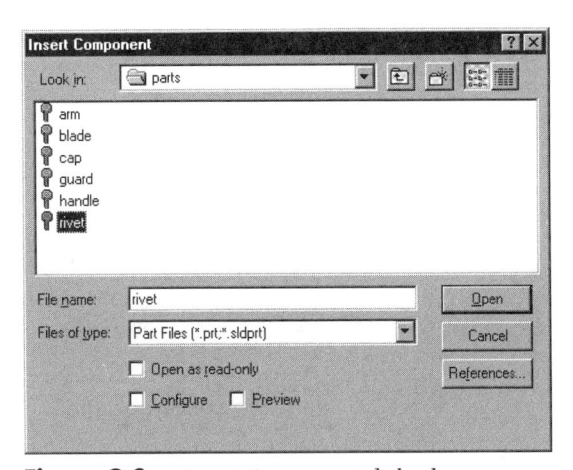

Figure 9.3. Insert Component dialog box

3. A part symbol appears next to the cursor, signifying that the selected part will be placed in the assembly. Bring the cursor over the origin. Two right-angled sets of arrows appear at the cursor. This means that the planes of the assembly and the planes of the rivet will be matched. When this symbol appears, click on the origin to place the part. The rivet appears with its origin coincident with the origin of the assembly and its planes coincident with the planes of the assembly, as shown in Figure 9.4. You may need to **Zoom In/Out** to enlarge the rivet.

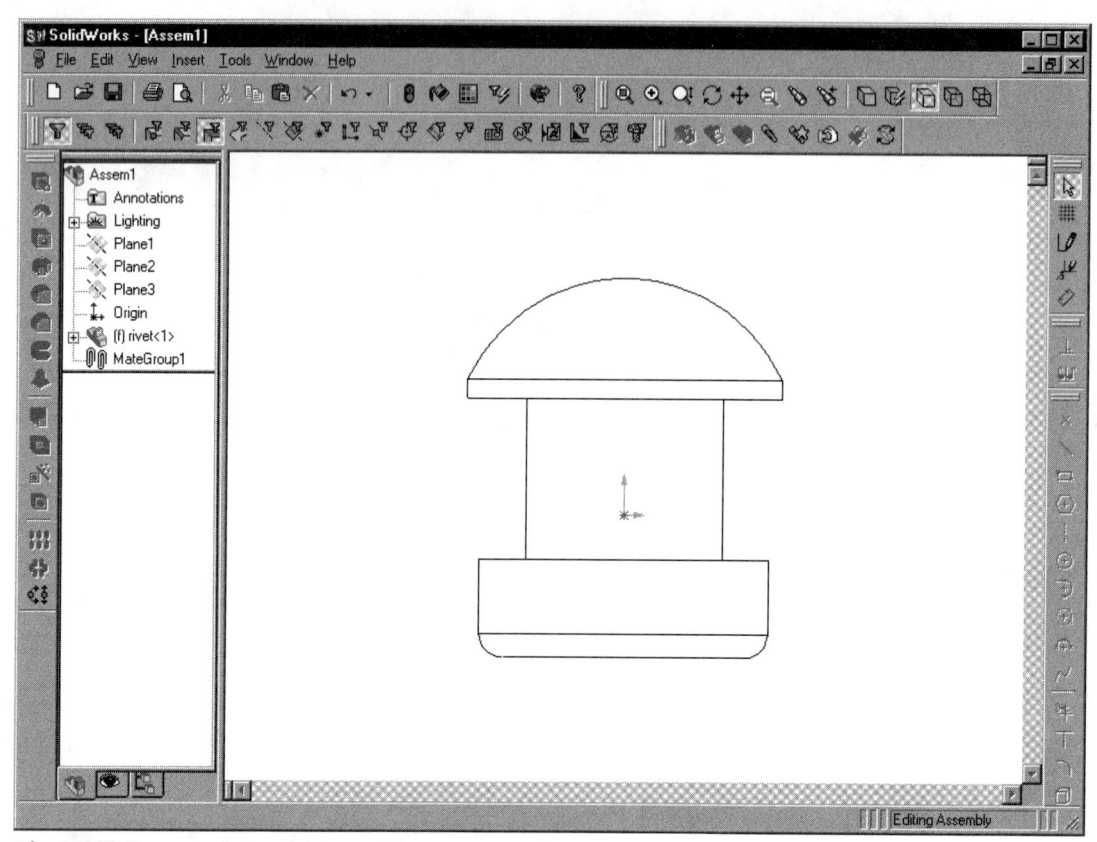

Figure 9.4. Rivet placed in assembly environment

4. Notice that **(f) rivet<1>** has appeared in the FeatureManager design tree. The "f" means that the rivet is fixed in the assembly space. The "1" means that this is the first instance of the rivet in the assembly. Click the plus sign next to **rivet** to show all of its features, just as when the part was being modeled. Click on the minus sign next to **rivet** in the FeatureManager design tree to hide the features of the rivet.

5. For clarity, uncheck **View ⇒ Origins**. This removes the arrows indicating the origin of the assembly.

9.1.3 Bringing the Arm into the Assembly and Orienting It

Now the second part, the arm, is brought into the assembly and oriented with respect to the rivet.

1. To bring the arm into the assembly, click **Insert ⇒ Component ⇒ From File**. Find the arm part and click **Open**.

2. Again, the cursor changes. In this assembly, only the rivet will be oriented with the origin of the assembly. Other parts will be oriented with respect to the rivet. Place the arm in the assembly by clicking anywhere on the screen. Notice that in the FeatureManager design tree **(-) arm<1>** appears. The minus sign indicates that the arm is free to move in space.

3. With the **Select** tool, click on any face of the arm. You may need to rotate the view. The selected face becomes outlined with a dashed blue line, and a box appears around the entire arm, signifying that the part is selected.

4. The **Assembly** toolbar, shown in Figure 9.5, is used to control the movement and placement of parts in an assembly. You can move the toolbar to another position by dragging the double bars at the end of the toolbar. The rightmost tools in the **Assembly** toolbar in Figure 9.5 are be used to move parts within an assembly:

   - The **Move Component** button (a hand holding a part) translates a part relative to the other parts in the assembly.
   - The **Rotate Component Around Axis** button (a part with an axis through it) rotates a part around a selected line or centerline.
   - The **Rotate Component Around Centerpoint** button (a part with rotation arrows around it) rotates a part around its own centerpoint, or origin. The **Move Component** and the **Rotate Component Around Centerpoint** are the two most useful tools for orienting parts. To use them, select a face of the part that you want to move. Next, click on the **Move Component** or **Rotate Component Around Centerpoint** button in the **Assembly** toolbar. Then click and drag to sorient the part. These two tools can also be found under **Tools** ⇒ **Component** ⇒ **Move** and **Tools** ⇒ **Component** ⇒ **Rotate**.

Figure 9.5. Assembly toolbar

5. Practice orienting the arm in various ways, using the **Move Component** and **Rotate Component Around Centerpoint** tools. Be sure that the arm is selected before moving it. If you select the rivet and try to move or rotate it, a dialog box alerts you that fixed components cannot be moved. Since the arm has not yet been fixed relative to the rivet, it can move and rotate freely. The PropertyManager window, which replaces the FeatureManager design tree when a part is moved or rotated, provides information about the position of the selected part. Orient the arm so that the hole in the arm is close to the rivet, similar to the orientation shown in Figure 9.6.

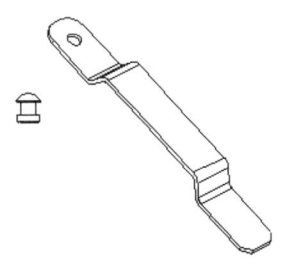

Figure 9.6. Arm close to rivet

9.1.4 Adding a Concentric Mate

After placing the first part, subsequent parts are oriented relative to other parts in the assembly using mates. For example, you will place the hole of the arm around the rivet with two mates: one concentric mate to match the hole's axis with the rivet's axis, and another mate to match the top face of the end of the arm with the underside face of the head of the rivet.

1. Click the **Zoom to Area** button in the **View** toolbar and zoom in on the hole and the rivet.

2. Select the cylindrical surface of the rivet using the **Select** tool. It will become highlighted. Add the cylindrical surface of the hole in the arm to the selection by *control clicking* the surface. Notice that it is easy to select surfaces because the **Selection Filter** is set so that only surfaces can be selected. These two surfaces, which will be mated, are shown in Figure 9.7.

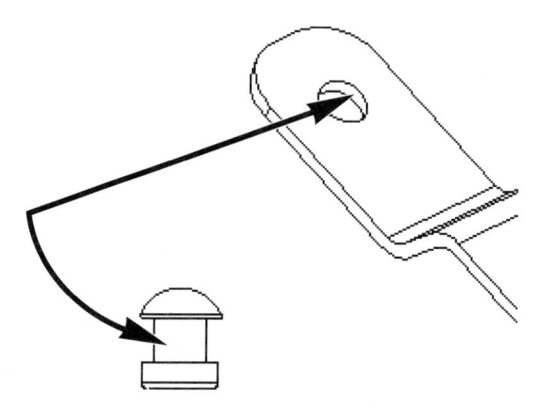

Figure 9.7. Arm and rivet surfaces

3. With the two surfaces selected, click the **Mate** button (a paper clip) in the **Assembly** toolbar, or select **Insert ⇒ Mate**.

4. The **Assembly Mating** dialog box appears, as shown in Figure 9.8. In the **Items Selected** field, the two faces that were selected are listed with their respective parts. In the **Mate Types** field, click the **Concentric** button. A concentric mate constrains the parts so that both circular features share the same axis of symmetry.

5. Click the **Preview** button to see what the assembly will look like after the mate is created. The screen should look similar to that shown in Figure 9.8 in the **Front** orientation. Note that the shaft of the rivet and the hole in the arm are aligned. It does not matter if the arm is above or below the rivet. If the arm is upside down, click the **Aligned** button in the **Alignment Condition** field to flip the arm and **Preview** again. Click **Apply** to create the mate.

6. In the FeatureManager design tree, click on the plus sign next to **MateGroup1**. The mate that was just created, **Concentric1 (rivet<1>, arm<1>)**, is listed. You may need to move the divider between the Solid-Works window and the FeatureManager design tree in order to see the entire

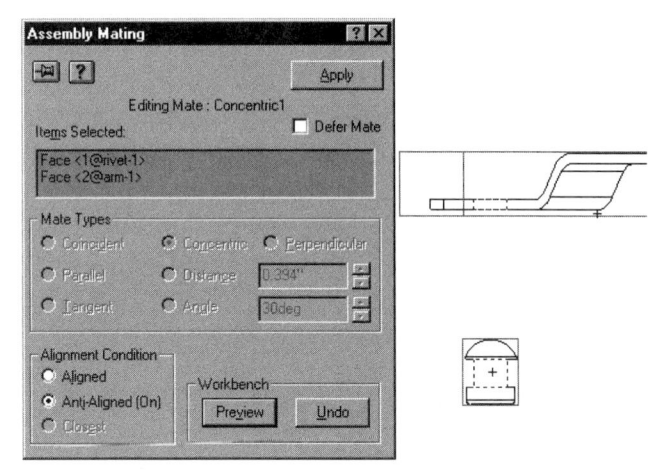

Figure 9.8. Assembly mating dialog box

name. To do this, move the cursor over the border between the FeatureManager design tree window and the Graphics Window until it becomes two vertical lines with horizontal arrows. Then, drag the border. As more and more mates are created, refer to the list under **MateGroup1** to keep track of them.

7. A part cannot be rotated around its centerpoint once it is mated, but it can be moved with the **Move Component** tool. **Zoom To Fit** and select the arm. Move it with the **Move Component** tool. The concentric mate constrains the hole in the arm to be concentric with the rivet. As a result, the arm can only be moved along the axis of the rivet or rotated around the rivet's axis. Open the ***Orientation*** dialog box if it is not already open, and click on the pushpin to keep it displayed. Set the view to ***Front*** by double clicking on ***Front***. Move the arm so that it is below the rivet.

9.1.5 Adding a Coincident Mate

The arm must be further constrained so that its top face is touching the underside face of the head of the rivet. This is called a *coincident* mate, since the plane of the arm's top face coincides with the plane of the underside of the rivet's head.

1. Use the **Rotate View** tool in the **View** toolbar to rotate the view so that the top face of the arm is visible.

2. Select the top face, as shown in Figure 9.9.

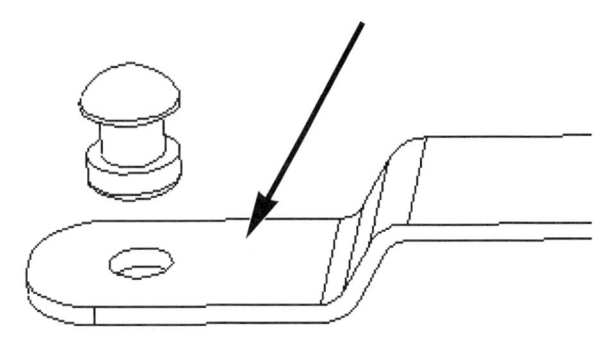

Figure 9.9. Arm surface

3. Reactivate the **Rotate View** tool and rotate the view so that the underside of the head of the rivet is visible. Click the **Rotate View** button again. This turns off the **Rotate View** tool and activates the **Select** tool. *Control click* on the face, as shown in Figure 9.10, to add it to the selection. Be sure that both the top face of the arm and the underside face of the rivet head are highlighted, to indicate both are selected. If not, move the cursor to an open space away from both parts, click, and start again.

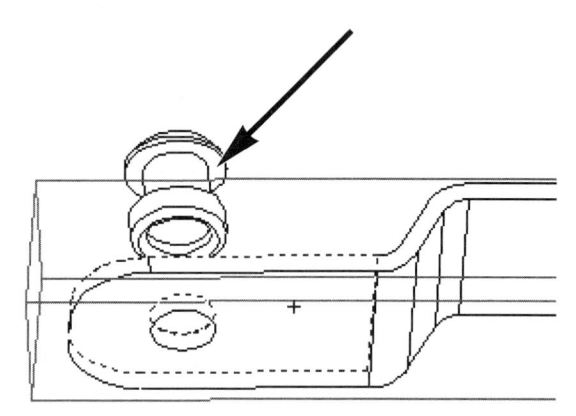

Figure 9.10.   Rivet surface

4. Click the **Mate** button in the **Assembly** toolbar. The face of the arm and the face of the rivet should be listed in the ***Items Selected*** field. Click the ***Coincident*** button. A coincident mate simply places two surfaces against each other, so that the surfaces are in the same plane.
5. ***Preview*** the mate. It might help to turn on **Shaded** display and rotate the view. If the mate looks wrong, click ***Undo***, close the ***Assembly Mating*** dialog box, and start over. Otherwise, click ***Apply*** and return to **Hidden Lines Removed**.

9.1.6   Assembling the Blade

The blade is added next. It will have two mates in the assembly. The hole of the blade is concentric with the cylindrical shaft of the rivet, and one face of the blade is coincident with the bottom face of the arm.

1. In the ***Orientation*** dialog box, set the view to ***Front***.
2. Bring the blade into the assembly and place it close to the other parts.
3. Select the surface of the blade. **Move** and **Rotate** the blade, so that you can see both the hole in the blade and the shaft of the rivet. Select both of these surfaces, as indicated in Figure 9.11. Be sure to *control click* the second surface.
4. With the two surfaces selected, create a concentric mate. It may help to **Redraw** the screen after the mate.
5. Click on the blade, and use the **Move Component** tool to move the blade below the arm and the rivet.

6. With the **Rotate View** tool, rotate the view so that the top surface of the blade can be seen. Then, rotate the view to see the bottom surface of the arm. If the blade obstructs the view of the arm, use the **Move Component** tool to

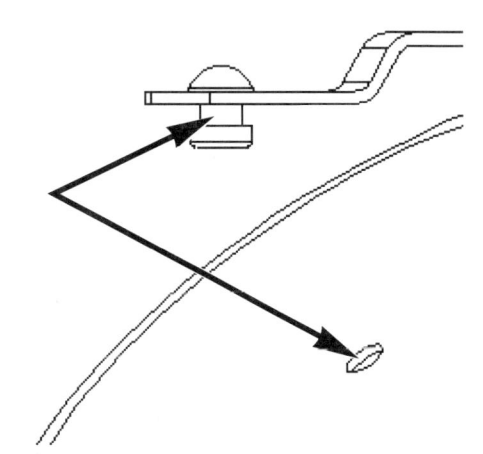

Figure 9.11. Blade and rivet selections

move the blade further below the arm. This will make it easier to select the two surfaces. Rotate the view and select the top surface of the blade.

7. Reactivate the **Rotate View** tool, and rotate the view so that the bottom face of the arm is visible. Click the **Rotate View** button again to turn off the **Rotate View** tool. *Control click* the bottom face of the arm. The two surfaces to be mated are shown in Figure 9.12.

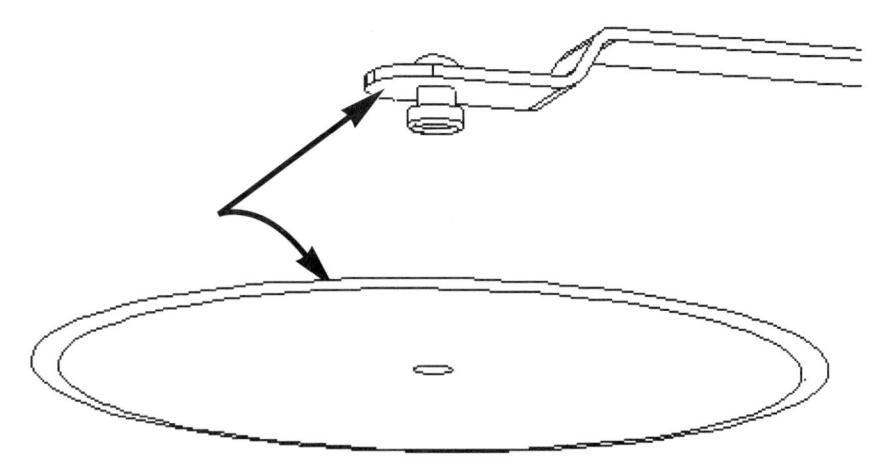

Figure 9.12. Blade and arm selections

8. Create a coincident mate between the selected surfaces and **Preview** the mate. It may help to turn on **Shaded** display and to rotate the view. Click ***Apply*** and return to **Hidden Lines Removed**.

9.1.7 Assembling the Second Arm

Although the assembly has two arms, only one part was modeled, because they are identical. Now a second arm is brought into the assembly and added to the other side of the blade.

1. Bring the arm into the assembly and place it below the existing components. Notice that a **<2>** appears next to **arm** in the FeatureManager design tree. This means that it is the second arm brought into the assembly.

2. Use the **Move Component** and **Rotate Component Around Centerpoint** tools to move and rotate the second arm. Place the second arm so that it closely mirrors the first arm and the hole is close to the rivet, as shown in Figure 9.13. It may help to use the **Shaded** display as you orient the arm. Then, return to **Hidden Lines Removed** display.

3. Apply a mate so that the hole of the arm is concentric with the shaft of the rivet. *Preview* the mate before clicking *Apply*. If the second arm does not mirror the first, click the **Aligned** or **Anti-Aligned** button in the **Alignment Condition** field to flip it. *Preview* the mate again and click *Apply*.

4. Select the bottom surface of the blade and the facing surface of the arm, as indicated in Figure 9.13. Create a coincident mate between the two selected surfaces.

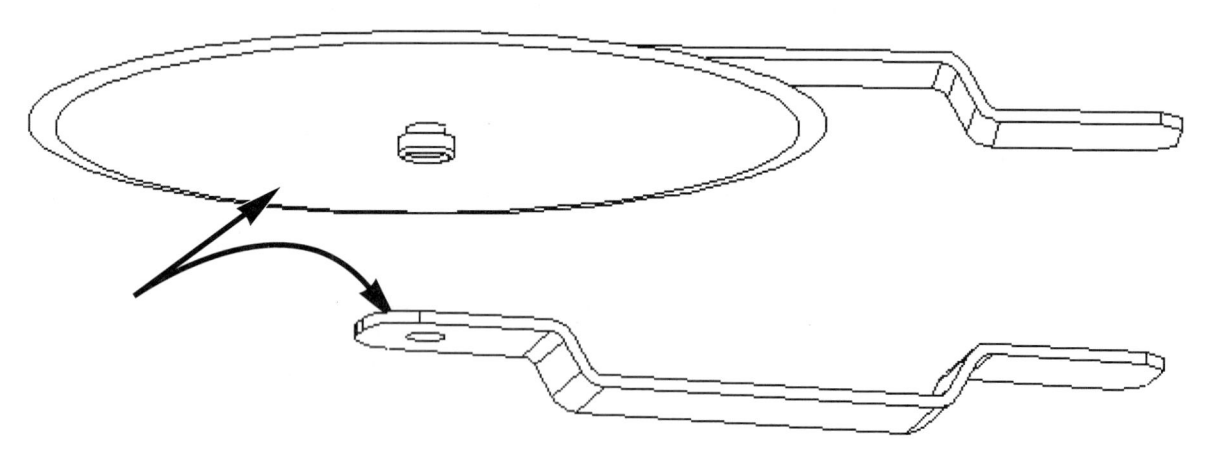

Figure 9.13. Arm2 and blade selections for the coincident mate

9.1.8 Adding a Parallel Mate

The arms can still rotate relative to each other, so they are probably not parallel to one another at this point. A parallel mate will constrain the arms so that they are aligned with each other.

1. Move one of the arms with the **Move Component** tool, so that they are approximately aligned with each other.

2. Select the side face (edge) of both arms, as indicated in Figure 9.14.

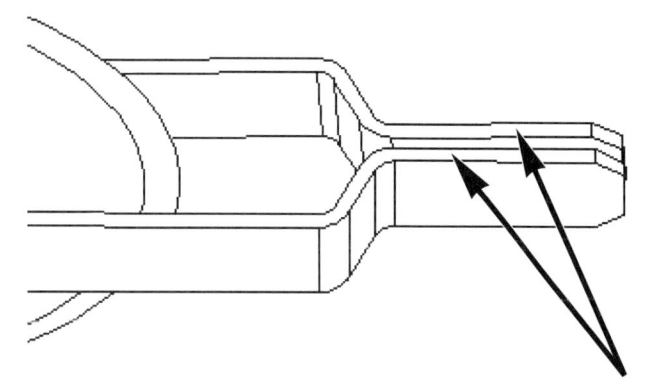

Figure 9.14. Arms parallel mate

3. Click the **Mate** button in the **Assembly** toolbar. In the **Assembly Mate** dialog box, click the **Parallel** button. This makes the two edge surfaces parallel to each other, removing any angle between the two arms.

4. When you are satisfied with the **Preview**, click **Apply** to create the mate.

5. With the parallel mate applied, one arm moves with the other. Try this using the **Move Component** tool.

Congratulations! This completes the cutter sub-assembly. The assembly should look like the one shown in Figure 9.1. **Save** the assembly as **cutter sub-assembly**. SolidWorks will ask you if you want to rebuild the assembly. Answer **Yes**, so that the changes that you have made will be implemented before saving. Since you will be using this sub-assembly in the next section, leave the **cutter sub-assembly** window open.

PROFESSIONAL SUCCESS: SMARTMATES

SmartMates allows components of an assembly to be mated without using the **Assembly Mating** dialog box to specify the type of mate. Many components that are mated together in assemblies are similar to each other. For example, some of the mates created in the pizza cutter are either a cylinder concentric with a hole or a face coincident with another face. The **SmartMates** feature recognizes these mates and others in order to allow for a quick assembly process.

To place a smart mate, click the **SmartMates** button in the **Assembly** toolbar. **Double click** the entity to be mated (e.g., a cylindrical section) on one component. After the cursor icon changes to a paper clip, click on the mating entity of another component (e.g., a circular hole). A mate is automatically created.

166 Chapter 9 Modeling an Assembly: The Pizza Cutter

9.2 MODELING THE PIZZA CUTTER ASSEMBLY

In this section, the cutter sub-assembly will be joined with the handle, the cap, and the guard to complete the pizza cutter assembly. The final pizza cutter assembly is shown in Figure 9.15.

Figure 9.15. Finished pizza cutter assembly

9.2.1 Creating a New Assembly and Inserting the Handle

A new assembly is created for the entire pizza cutter.

1. Click the **New** toolbar button or **File** ⇒ **New** to model a new assembly. Select *Assembly*.
2. Bring the handle into the assembly by clicking **Insert** ⇒ **Component** ⇒ **From File**. Remember to set the *Files of Type* pull-down menu to *Part Files*. Bring the cursor over the origin so that the two right-angled sets of arrows appear. Place the handle at the origin. This orients the planes of the handle with the planes of the assembly, so that the origins coincide.
3. Be sure that the **Filter Faces** toolbar button is selected, so that only faces of parts can be selected. Turn off **Origins** in the **View** menu to hide the origin. Set the view to **Hidden Lines Removed**. Now you are ready to begin assembling the rest of the pizza cutter.

9.2.2 Assembling the Cap

The next part added to the assembly is the cap. It is placed on the end of the handle with the reduced diameter.

1. Bring the cap into the assembly and place it near the bottom of the handle.

2. Select the cap and rotate it using **Rotate Component Around Centerpoint**, so that the concave side faces the bottom of the handle.
3. Select the two surfaces, as shown in Figure 9.16.

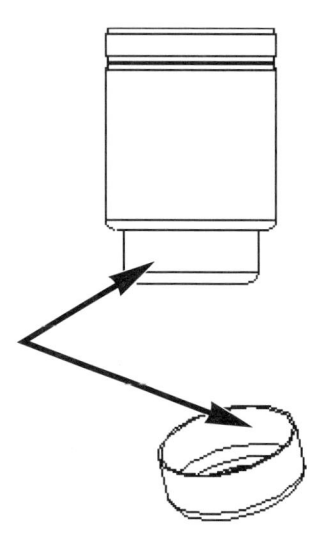

Figure 9.16. Handle and cap selections

4. Click the **Mate** button in the **Assembly** toolbar, or click **Insert ⇒ Mate**, to mate the two cylindrical surfaces.
5. In the **Assembly Mate** dialog box, click the **Concentric** button. If the **Preview** shows the cap facing the wrong direction, click **Aligned** in the **Alignment Condition** field to flip the cap. Click **Apply** to create the mate. Check the constraint by selecting the cap and using the **Move Component** tool. The cap should only move along the axis of the handle.
6. A second constraint is needed to position the cap at the end of the handle. Select the inside flat surface of the cap and the bottom surface at the end of the handle. You will need to rotate the view to select both surfaces. The inside surface of the cap that should be selected is shown in Figure 9.17.

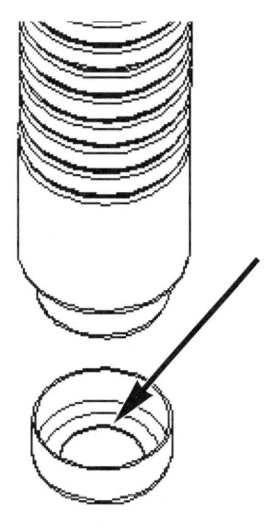

Figure 9.17. Cap: coincident selection

7. Create a coincident mate between the two surfaces in order to bring the inside of the cap up to the bottom surface at the end of the handle. Switch to **Shaded** view to see the cap on the handle clearly. If the preview appears to be correct, click **Apply**. Then, return to **Hidden Lines Removed**.

9.2.3 Assembling the Guard

The guard is placed on the bottom face of the cap, and aligned so the axis of the hole in the guard coincides with the axis of the handle.

1. Insert the guard into the assembly and place it near the cap.

2. Mate the surface of the guard to the bottom surface of the cap using a coincident mate. Be sure to select the surface of the cap, rather than the handle. When the **Select** tool is over the correct surface of the cap, a box should pop up near the cursor, indicating **Base-Revolve-Thin of Cap<1>**. The guard may not actually touch the cap after the mate has been created. The planes of the surfaces, which have infinite length, have been mated. To see this, go to the **Front** view. Then, move the guard from side to side. Although it can move left or right, it will always remain so that the lower surface of the cap is coincident with the plane of the upper surface of the guard, as shown in Figure 9.18.

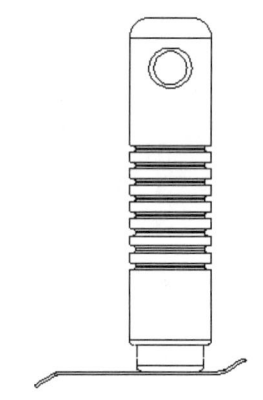

Figure 9.18.   Guard mated to handle

3. To fully constrain the position of the guard, you can match the planes of the guard with the fixed planes of the assembly. If necessary, click on the leftmost tab, as shown at the bottom of Figure 9.19, to display the FeatureManager design tree. Click on the plus sign next to **guard**. The FeatureManager design tree should look similar to the one shown in Figure 9.19. The features of the guard are displayed, including its three planes. Click on each of the planes to see where they are. Similarly, click on each of the planes of the assembly in the FeatureManager design tree. You will be mating **Plane1** and **Plane3** of the guard to **Plane1** and **Plane3** of the assembly, respectively.

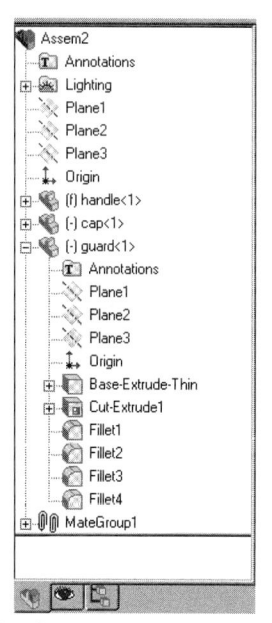

Figure 9.19. Feature manager design tree

4. Click on **Plane1** of the **guard** in the FeatureManager design tree. The plane becomes highlighted in green.
5. *Control click* **Plane1** of the assembly in the FeatureManager design tree (under **Assem2** in Figure 9.19) to add it to the selection.
6. Click the **Mate** button in the **Assembly** toolbar. Both planes should be in the ***Items Selected*** list. Set the mate type to be a ***Coincident*** mate and click **Apply**. This aligns **Plane1** of the guard to **Plane1** of the assembly.
7. Mate **Plane3** of the guard to **Plane3** of the assembly in the same way. This completely constrains the position of the **guard** relative to the handle. Click the minus sign next to guard in the FeatureManager design tree.

9.2.4 Hiding an Object

The next component to be added to the assembly is the cutter sub-assembly. The arms of the cutter sub-assembly will be inserted into the rectangular hole at the end of the handle. Since the guard partially covers the view of the hole, you will hide it temporarily, so that the mate between the handle and the cutter sub-assembly can be created more easily.

1. *Right click* **guard** in the FeatureManager design tree.
2. Select **Hide Component** from the menu that appears. The guard is now hidden. The outlined part icon to the left of **guard** in the FeatureManager design tree indicates that the guard is not shown on the screen.

9.2.5 Inserting the Sub-assembly into the Assembly

A sub-assembly can be mated, moved, and rotated just like any part in an assembly. Here, the cutter sub-assembly will be added to the main assembly.

1. Maximize the SolidWorks window by clicking the **Maximize** button in the upper right corner of the Graphics Window, to provide enough space for the assembly. Then, select **Window** ⇒ **Tile Horizontally** to see all of the active windows. If the cutter sub-assembly has been closed, click **File** ⇒ **Open** to open it. Then, select **Window** ⇒ **Tile Horizontally** again. The screen should look similar to that shown in Figure 9.20.

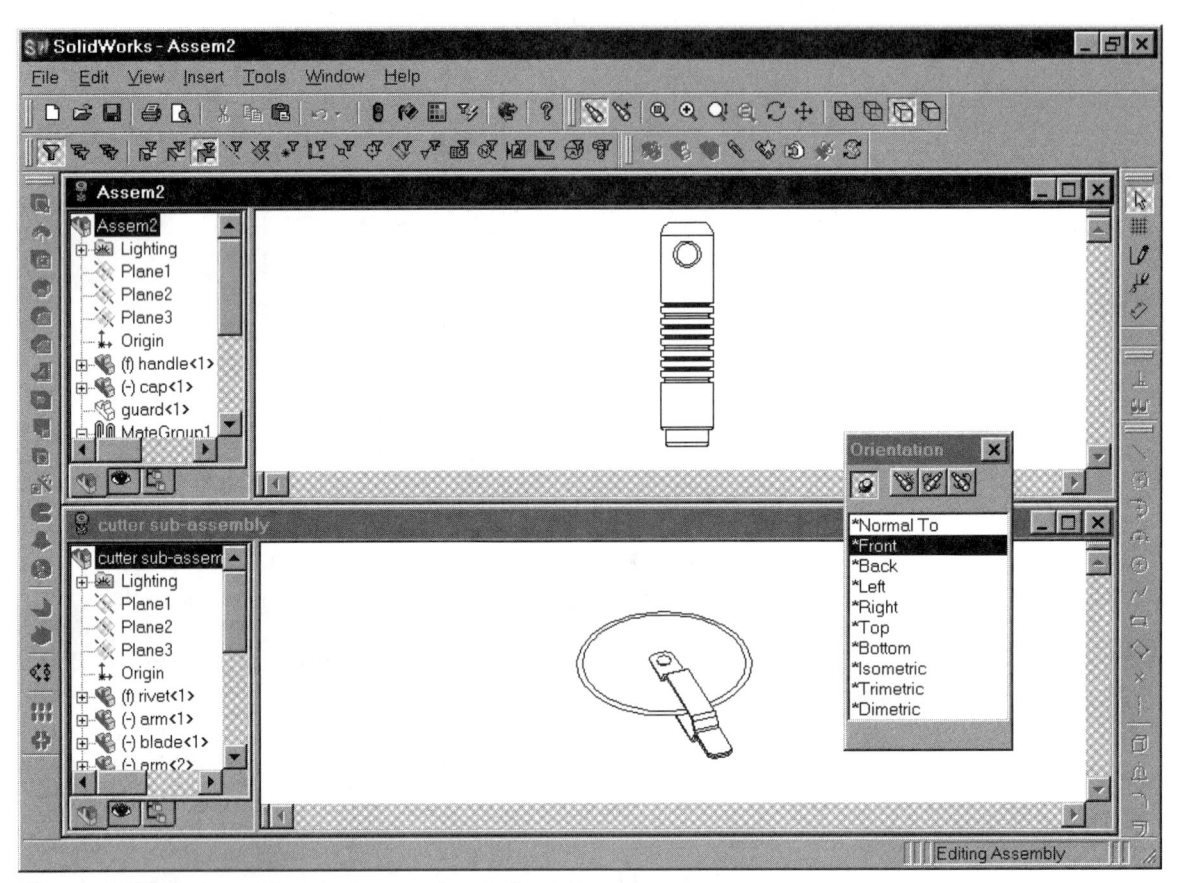

Figure 9.20. Assembly and sub–assembly windows

2. In the FeatureManager design tree of the cutter sub-assembly window, click and drag **cutter sub-assembly** to the main assembly window (above the sub-assembly window). The component icon appears next to the cursor, signifying that a component is being brought into the assembly. Release the mouse button to place the cutter sub-assembly in the main assembly window.

3. Maximize the main assembly window, so that you have more room to work. If necessary, **Zoom to Fit** to show both the cutter sub-assembly and the handle.

4. Move and rotate the cutter sub-assembly, so that the ends of the arms are close to the rectangular hole at the end of the handle. Be sure that the **Filter Faces** toolbar button is selected.

5. Zoom in as shown in Figure 9.21. Select the flat face of one of the arms and the long side of the inside of the rectangular hole of the handle. You will need to rotate the view to select both surfaces. To ensure a correct mate, the surfaces should be facing each other.

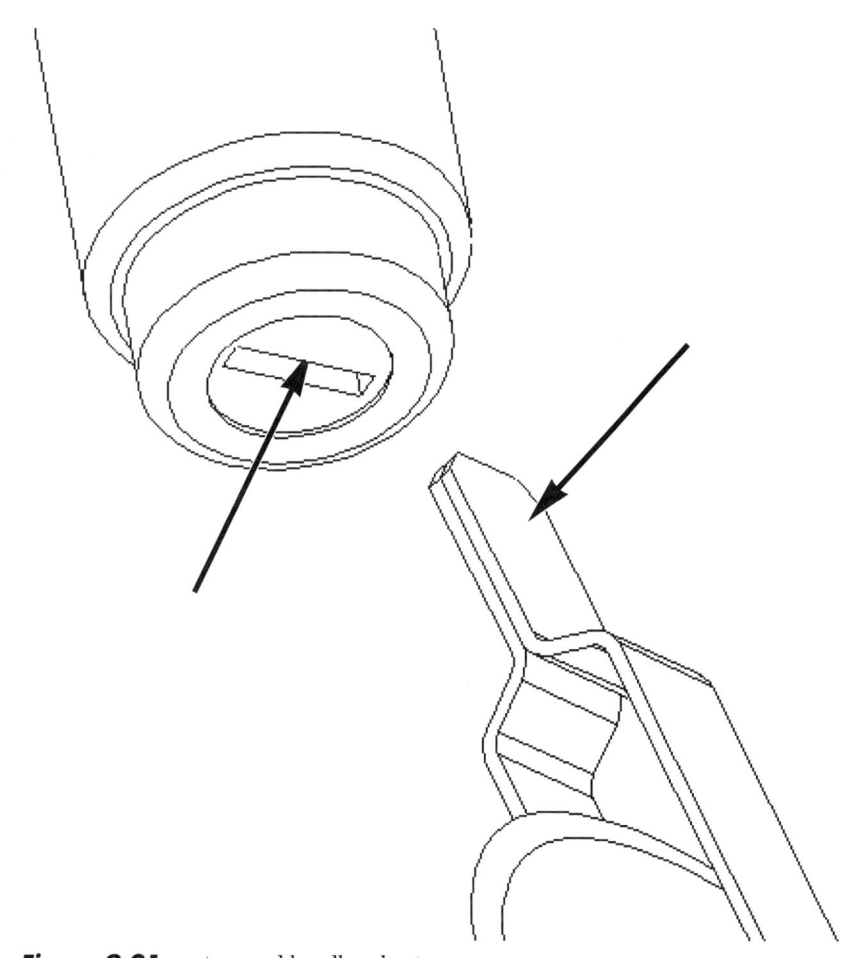

Figure 9.21. Arm and handle selections

6. Use a coincident mate to make the two faces coplanar. While previewing the mate, go to the **Left** view and turn on **Hidden In Gray** to see the hidden lines. If the arm looks like it will interfere with the handle instead of fitting in the hole, flip the cutter sub-assembly by clicking *Aligned* or *Anti-Aligned* in the **Assembly Mate** dialog box. Click *Apply*. Return to **Hidden Lines Removed**.

7. Rotate the view and zoom as necessary to see the short side of the inside surface of the rectangular hole in the handle. Select the short side, as shown in Figure 9.22. Rotate the view and add the edge of the arm to the selection, as shown in Figure 9.22.

8. Add a coincident mate between these two surfaces so that they are coplanar. ***Preview*** the position in the **Front** view. Click ***Apply***.

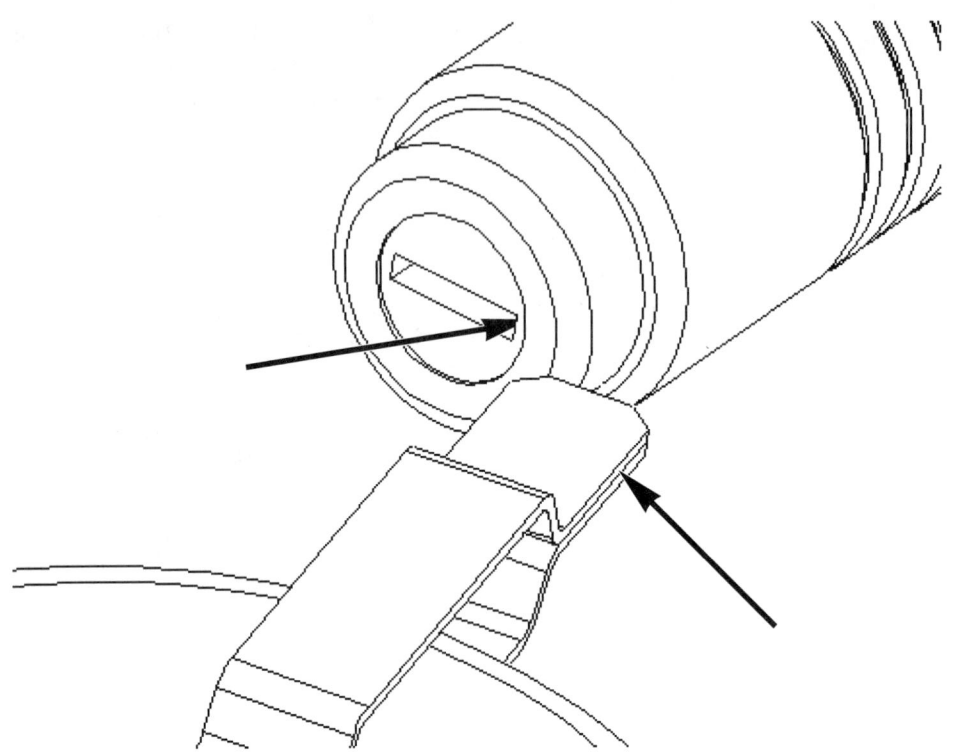

Figure 9.22. Arm and handle: small side selections

9.2.6 Showing the Guard and Finishing the Assembly

The cutter sub-assembly can still move along the axis of the handle. To fully constrain the sub-assembly, it will be mated to the guard.

1. Make the guard visible by *right clicking* on **guard** in the FeatureManager design tree and selecting **Show Component**. The guard reappears.

2. In the following steps, you will mate an edge of the arm to the underside face of the guard. Rotate the view of the assembly, zoom in, and select the face of the guard closest to the cutter sub-assembly, as shown in Figure 9.23.

3. Since an edge will be selected next, turn on the **Filter Edges** button and turn off the **Filter Faces** button in the **Selection Filter** toolbar. Add the edge to the selection as shown in Figure 9.23. This is the edge where the curved corner meets the flat face of the arm. If you cannot see the edge, be sure that **View ⇒ Display ⇒ Tangent Edges Visible** is checked.

4. Click the **Mate** button in the **Assembly** toolbar and add a coincident mate. This makes the line forming the edge of the arm coincident with the plane of the face of the guard. In other words, the line of the edge of the arm is constrained to be in the plane of the face of the guard. Click *Preview* and then *Apply*. The assembly should look like Figure 9.24.

5. Because you have finished creating the assembly, you will not need the **Selection Filter** toolbar from now on. First, click the **Clear All Filters** toolbar button. Then, remove the toolbar by unchecking Selection **Filter** in the **View ⇒ Toolbars** menu.

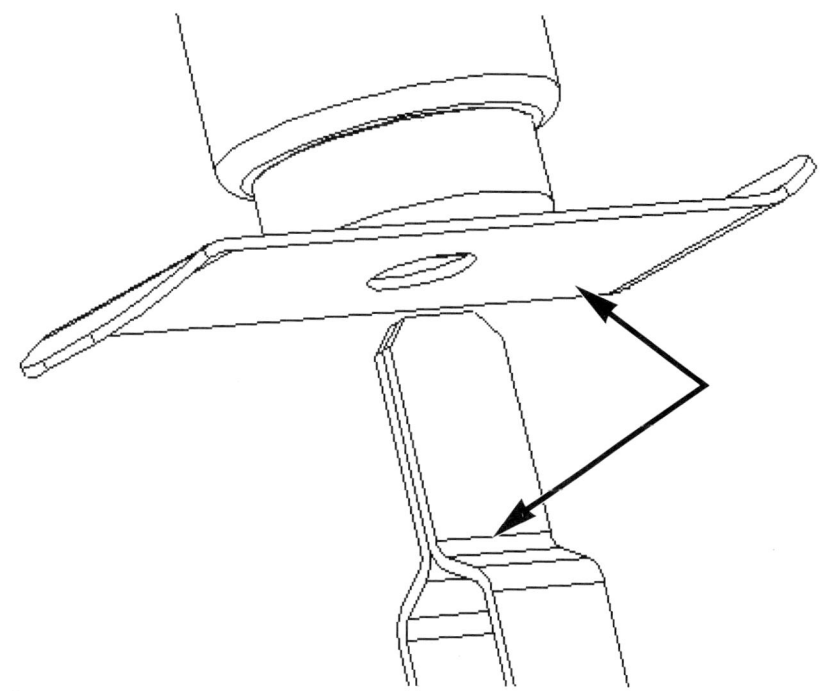

Section 9.2 Modeling the Pizza Cutter Assembly **173**

Figure 9.23. Arm and guard: coincident mate selections

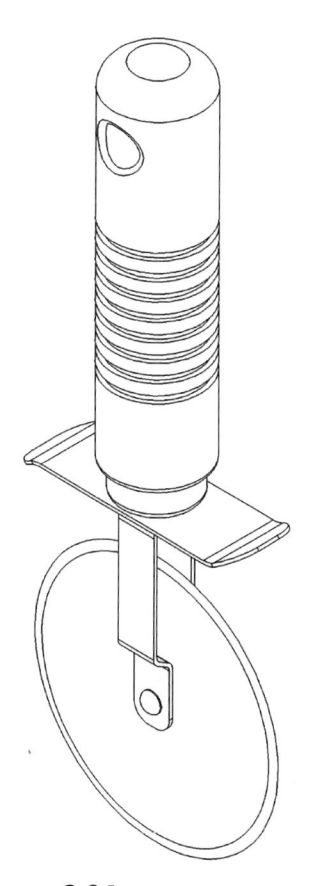

Figure 9.24. Finished assembly

Congratulations! This completes the assembly. Rebuild your assembly by clicking the **Rebuild** button (a stoplight) in the **Standard** toolbar, or **Edit** ⇒ **Rebuild**. Save your work as *pizza cutter*. Leave the **pizza cutter** window open for the next section, in which you check the assembly for interference. Close the **cutter sub-assembly** window by clicking on it in the **Window** menu, followed by **File** ⇒ **Close**.

9.3 CHECKING THE ASSEMBLY FOR INTERFERENCE

Now that you have successfully assembled the pizza cutter, the next task is to determine whether the parts fit together properly. SolidWorks simply assembles the parts where the placement is specified. It does not immediately check to determine whether any part interferes with another part. However, SolidWorks has built-in functionality to check for these types of design errors. This is done by comparing the solid volumes and placements of every part in the assembly to find interference between components.

9.3.1 Checking for Interference Volumes

The interference detection feature of SolidWorks will be used to determine whether any of the components of the pizza cutter interfere.

1. Be sure the **pizza cutter** window is open. Click on **Tools** ⇒ **Interference Detection**. The *Interference Volumes* dialog box appears.

2. The pizza cutter assembly appears in the **Selected Components** list. The position of each part in the assembly will be checked in relation to its neighboring components.

3. Click the **Check** button to have SolidWorks check for interference. SolidWorks generates a list of the interferences in the assembly. *Interference1* and *Interference2* should appear under *Interference results* box. Change the view to **Hidden In Gray**, and zoom in to the cap of the pizza cutter. Click on each item in the *Interference results* list to see the location of each highlighted interference on the assembly, as shown in Figure 9.25.

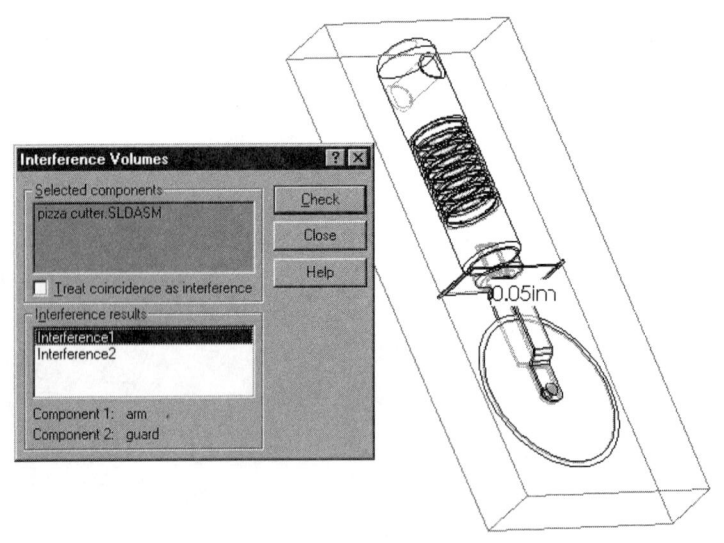

Figure 9.25. Interference Volumes dialog box

When you click on any interference on the **Interference results** list, its volume is highlighted in blue on the assembly. The components that are interfering are shown next to **Component 1:** and **Component 2:** just below the **Interference Results** box. In this case, each interference is between the guard and one of the arms. A close up of one of the interferences is shown in Figure 9.26. Click the **Redraw** button to remove the number and show the interfering volume in yellow.

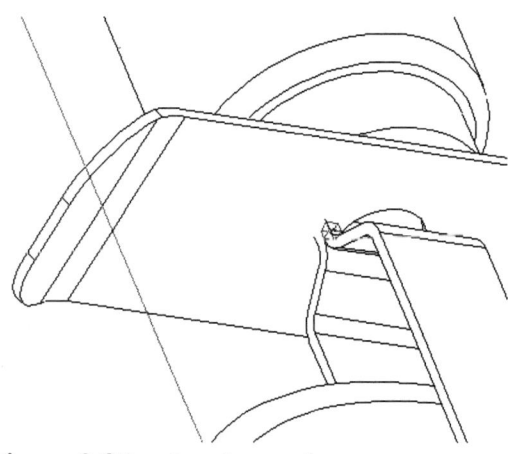

Figure 9.26. Interference closeup

After clicking each interference, it becomes evident that the rectangular arms are too wide to fit through the circular hole in the guard. To remedy this, you will reopen the guard and change the circular hole to a rectangular hole. The assembly will be rebuilt, once the guard is modified to incorporate the redesigned hole. To continue, close the **Interference Volumes** dialog box by clicking **Close**.

9.3.2 Opening a Sketch Within the Guard

Two options are available to edit the geometry of the guard. The first option, **Edit Part**, permits changes in the properties of the part while still in the assembly window. This is useful for changing the values of dimensions and making other minor changes. The second option, **Open guard.sldprt**, opens the **guard** file in a new window. It is easier to make major changes to an individual part using this option.

1. Bring the mouse over the guard in the Graphics Window and *right click* it. Select **Open guard.sldprt** from the menu. The guard opens in its own window. If you click on the **Window** menu, you can see that the pizza cutter assembly is still available. The assembly window is behind the **guard** window.

2. The features that were modeled in the guard are listed in the FeatureManager design tree. **Cut-Extrude1** represents the hole. Click on the plus sign to the left of **Cut-Extrude1**. **Sketch2** is the sketch of the circular hole. This sketch can be opened and revised, so that a rectangle replaces the circle. When the sketch is exited, the model of the guard will be "rebuilt", and the hole will be rectangular.

3. *Right click* **Sketch2** and select **Edit Sketch** from the menu. This opens the sketch. *Double click* **Normal To** in the **Orientation** dialog box to view the sketch normal to the guard's face.

9.3.3 Creating the Rectangular Hole

Now that the circular cut is visible in its sketch plane, you can change the hole to a rectangular one.

1. With the **Select** tool, click on the circle. It will become highlighted in green. Hit the delete key on the keyboard, or **Edit ⇒ Delete** to remove the circle. The dialog box shown in Figure 9.27 appears. This dialog box is referring to the dimension on the circle which will be removed when the circle is deleted. You need to delete this dimension, so click **Yes**. The circle and the dimension are removed from the sketch.

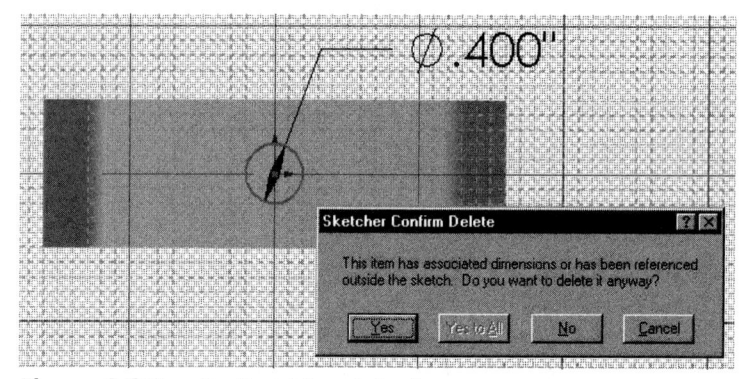

Figure 9.27. Confirm delete dialog box

2. Activate the **Rectangle** tool in the **Sketch Tools** toolbar, or **Tools ⇒ Sketch Entity ⇒ Rectangle**. Draw a rectangle that surrounds the origin by clicking to the bottom left of the origin and dragging the rectangle above and to the right of the origin.

3. Place the four dimensions, as shown in Figure 9.28. These dimensions describe the size of the rectangle and locate the rectangle relative to the origin. Your dimensions will likely be different. Do not edit the numerical dimensions to match those in Figure 9.28.

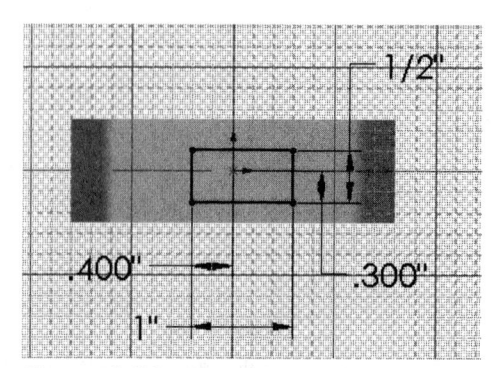

Figure 9.28. Rectangle hole sketch

9.3.4 Adding Equations to Dimensions

When dimensioning, it is sometimes useful to describe dimensions using *equations* rather than entering numeric values. For example, one dimension can be set to be two times the value of another dimension. If the first dimension ever changes, the second dimension will be updated accordingly. In this sketch, you will add equations to the rectangle's dimensions to ensure that the rectangle is always centered about the origin.

1. Activate the **Select** tool. Then, click **Tools** ⇒ **Equations**. The *Equations* dialog box appears.

2. Click on the numerical value for vertical dimension, which happens to be .300 in the sketch shown in Figure 9.29. Click *Add* in the **Equations** dialog box to add the dimension to the new equation. *"D3@Sketch2"* or something similar appears in the *New Equation* dialog box, as shown in Figure 9.29. In the example that is shown, the "3" means it is the third dimension placed in **Sketch2**. Since the "3" reflects the order in which the dimensions were placed, your number may be different.

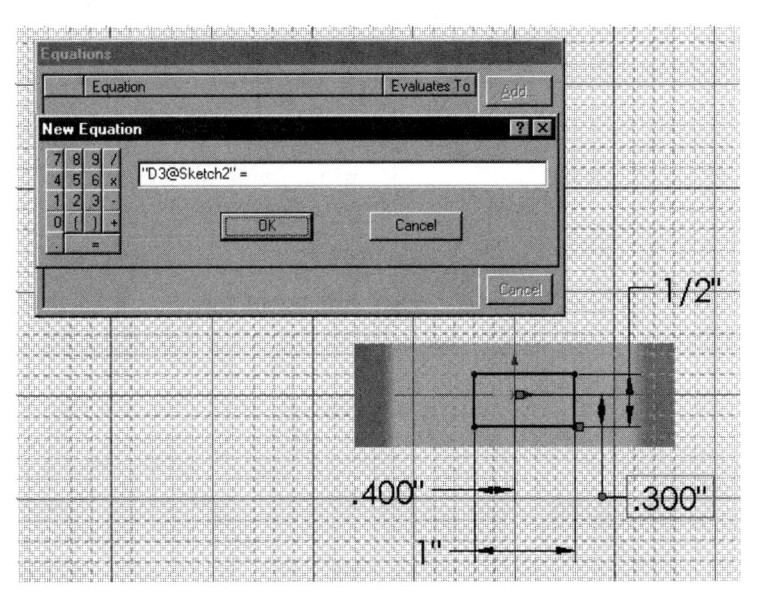

Figure 9.29. Equation dialog box

3. Click on the other vertical dimension in the sketch, which happens to be 1/2 in Figure 9.29. This adds *"D1@Sketch2"* to the equation. (Again, your number may be different.)

4. Type in *"/2"* (without quotes) at the end of the equation by using the keyboard or by clicking on items on the keypad in the *New Equation* dialog box. This sets the smaller vertical dimension to be half of the larger vertical dimension. The entire string should look something like this:

"D#@Sketch2"="D#@Sketch2"/2

The pound symbols represent the dimension numbers specific to your sketch.

5. Click **OK**. The equation is listed in the dialog box, and the value of the smaller vertical dimension is shown in the ***Evaluates To*** column. Be sure that the smaller dimension is half the value of the larger vertical dimension. The dimension on the sketch will not yet update. If you need to edit an equation, click the ***Edit All*** button.

6. Repeat Steps 2 through 5 above to create an equation that makes the value of shorter horizontal dimension half of the longer horizontal dimension. Click **OK** in the ***Equations*** dialog box when you are done. Now, the dimensions on the sketch update to the values prescribed by the equations.

7. Since there are two "free" dimensions and two dimensions with which equations are associated, you only need to define two of the four dimensions in this sketch. With the **Select** tool, *double click* the larger vertical dimension. Set the value to **.096** and click on the green check mark. The smaller vertical dimension value should automatically update to **.048**.

8. Set the larger horizontal dimension to be **1/2**. This sets the value of the smaller horizontal dimension to **1/4**. The rectangle should be centered at the origin. Note that if you attempt to adjust a dimension driven by an equation, SolidWorks will warn you that the value cannot be changed.

9. Exit the sketch by clicking on the **Sketch** button in the **Sketch** toolbar. Because **Cut-Extrude1** is defined to cut the sketch through the part (regardless of the shape of the cut), the guard rebuilds with a rectangular hole.

10. Before returning to the assembly, save the guard with the same name by selecting **File ⇒ Save** or using the **Save** toolbar button.

9.3.5 Rebuilding the Pizza Cutter Assembly with the New Guard

Now you can rebuild the pizza cutter assembly with the modified guard.

1. Select **pizza cutter** from the **Window** menu to return to the assembly. The dialog box shown in Figure 9.30 appears, asking if the model should be rebuilt or updated, to reflect the changes in the guard. The assembly should be updated with the modified guard, so click **Yes**.

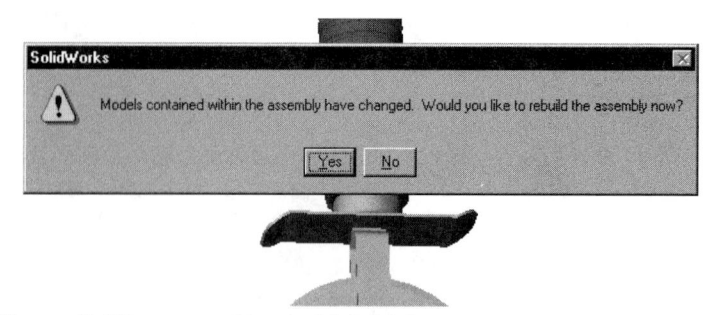

Figure 9.30. Assembly rebuild dialog box

2. The assembly updates the guard component to reflect the changes that were made. Rotate the assembly to be sure that the rectangular hole is in the right position with respect to the handle. Change the display to **Hidden In Gray**,

if necessary. To see the rectangular hole in the guard more easily, hide the cutter sub-assembly by *right clicking* on it in the FeatureManager design tree and selecting **Hide Component**. When satisfied that the rectangular holes in the guard and the handle align, *right click* on the cutter sub-assembly in the FeatureManager design tree, and select **Show Component**.

3. Now, check again to be sure that there is no interference within the assembly. Click on **pizza cutter** in the FeatureManager design tree to select it. Select **Tools ⇒ Interference Detection**. The pizza cutter assembly should be listed in the *Selected Components* field. Click the **Check** button. The result should be *0 Interference*, indicating no interference. Close the *Interference Volumes* dialog box.

4. Save the pizza cutter assembly. If a dialog box appears about saving the referenced models, click **Yes**, so that the updated models are saved. In the **Window** menu, click on **guard** and **Close** the guard window.

Congratulations! The final pizza cutter assembly should look like Figure 9.15. You have successfully assembled the pizza cutter and resolved interference errors.

9.4 CREATING AN EXPLODED VIEW OF THE ASSEMBLY

Exploded views are helpful to see how parts the assembly fit together and to visualize the assembly procedure. Each component is separated from the components with which it mates, usually along the assembly axis, as shown in Figure 9.31. The pizza cutter can be returned to its assembled configuration, once the exploded view is created.

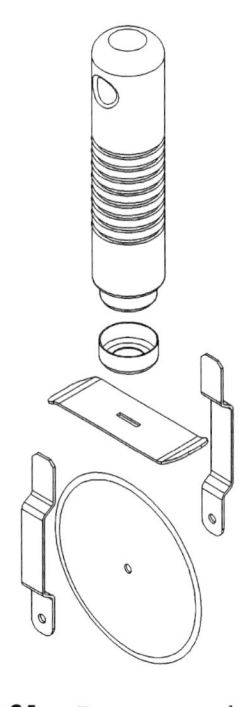

Figure 9.31. Pizza cutter exploded

9.4.1 Exploding the Cutter Sub-assembly from the Handle

First the cutter sub-assembly is exploded from the handle. Later, the rest of the assembly will be exploded.

1. Before exploding the assembly, **Zoom To Fit** and use **Hidden Lines Removed** display. Click anywhere on the background in the Graphics Window, so that nothing is selected.

2. Select **Insert ⇒ Exploded View**. The *Assembly Exploder* dialog box shown in Figure 9.32 appears. To see the full dialog box, which is shown in Figure 9.32, click on the *New* button (a staircase). This opens a new step of the exploded view. One or more components are exploded in each step. An exploded view can contain several steps.

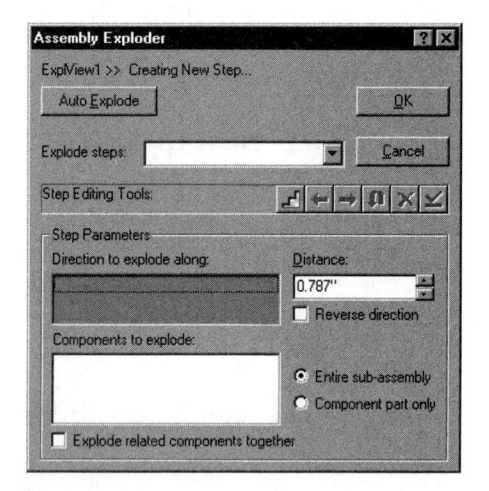

Figure 9.32. Assembly Exploder dialog box

3. The *Assembly Exploder* dialog box guides you through each stage of a step by highlighting (in pink) the current field that requires an input. The *Direction to explode along* field should be highlighted now. If it is not, click on this field. Use the **Front** view of the assembly. Click on the shaft of the handle. An arrow appears, indicating the direction of the explosion, as shown in Figure 9.33. **Face of handle<1>** appears in the field.

4. The *Components to explode* field is now highlighted. Click on the blade to select the cutter sub-assembly. *cutter sub-assembly<1>/blade<1>* appears in the field. Leave the *Entire sub-assembly* button checked, so that the entire sub-assembly will be exploded.

5. In the *Distance* field, enter 3. Click *Apply* (the green check mark) in the *Assembly Exploder* dialog box in order to see the explosion step. If the sub-assembly exploded in the wrong direction, click *Reverse Direction*. *Double click Isometric* in the *Orientation* dialog box. The screen should look similar to that shown in Figure 9.34.

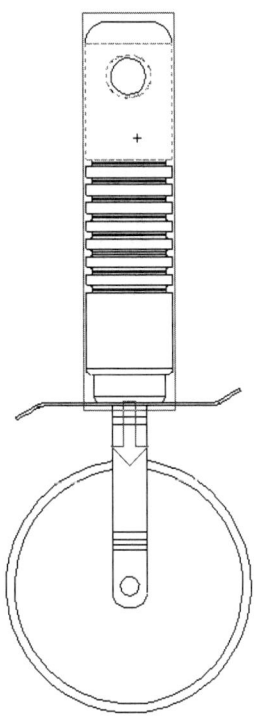

Section 9.4 Creating an Exploded View of the Assembly **181**

Figure 9.33.   Cutter explode direction

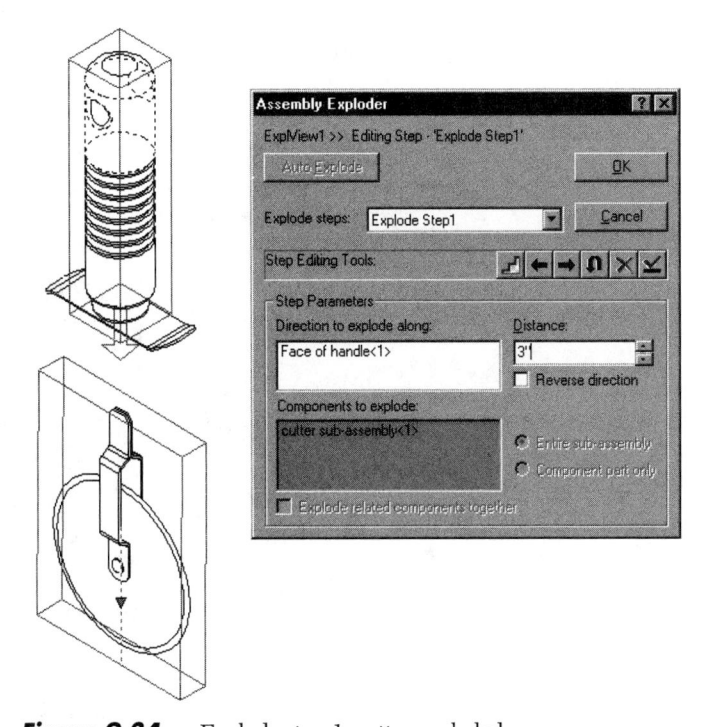

Figure 9.34.   Explode step 1, cutter exploded

9.4.2 Exploding an Arm

Next, you will explode one of the arms in the cutter sub-assembly.

1. Click the **New** button (a staircase) in the **Assembly Exploder** dialog box to start a new step.
2. To have the arms explode outward away from the blade, a direction to explode along must be defined. You can use the circular hole at the top of the handle, as shown in Figure 9.35, to indicate the direction for the arm to be exploded. If necessary, move the dialog box to see the hole. Select the inside surface of the circular hole. The face should be listed in the **Direction to explode along** field. The arrow should be pointing along the axis of the hole.

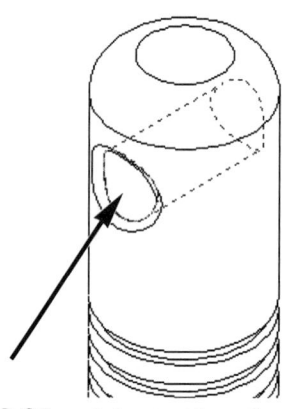

Figure 9.35. Select inside surface of hole

3. Click on the **Component part only** button so that only the arm will be exploded, rather than the entire cutter sub-assembly. Click on the arm to be exploded in the direction indicated by the arrow at the top of the handle, as shown in Figure 9.36. This adds the arm to the **Components to explode** field. This is probably the arm on the near side of the blade.
4. Set the **Distance** to **2** and click the **Apply** button (a green check mark). The arm moves outward from the blade.

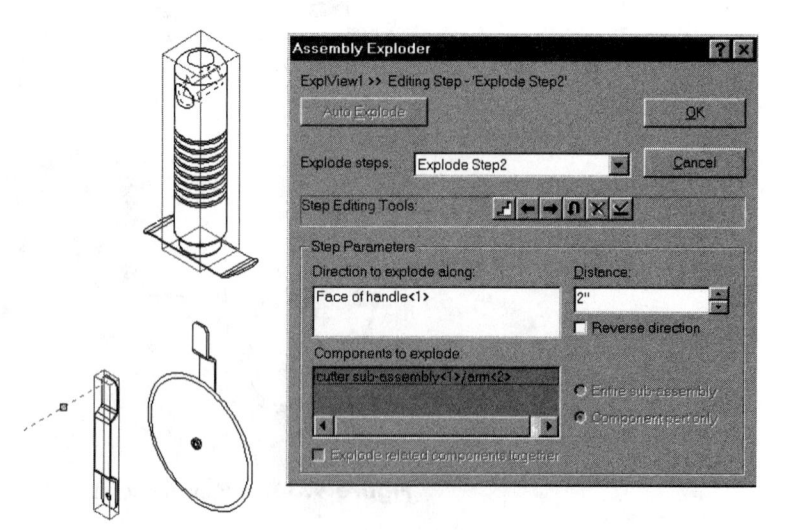

Figure 9.36. Explode step 2, arm exploded

9.4.3 Exploding the Other Components

The other components can be exploded in a similar manner. Before each step, click the *New* button to start a new explosion step. Use the *Delete* button (a red X) in the *Assembly Exploder* if you make a mistake. There will be six explode steps in all. Two have already been completed.

1. Explode the other arm in the other direction away from the blade. Use the same surface of the circular hole in the handle used to explode the first arm, but be sure to check the *Reverse Direction* ckeckbox in order to explode the second arm to the opposite side. Also, be sure to click the *Component part only* button. Use a distance of **2** inches.
2. Explode the rivet **4** inches to the side of the blade toward the head of the rivet. It may be necessary to zoom in to see which end of the rivet is on the near side of the blade. Use the inside surface of the circular hole in the handle as the *Direction to explode along*. Since the exploded view indicates assembly direction, the rivet should be exploded to the side of the assembly that the rivet's head is on. Click the *Component part only* button, then select the rivet.
3. Explode the guard downward **2.5** inches using the handle as the direction to explode along.
4. Explode the cap downward **1.25** inches.
5. Note that the explode steps are listed in the *Explode steps* pull-down menu. Clicking on any particular step in the menu causes that step to be highlighted. Each step can be modified. For instance, the distance that a part is exploded can be changed. When the *Apply* button is clicked, the modifications are applied. Make any necessary adjustments to the exploded view. When you are satisfied with the results, click *OK*. Click anywhere on the background, so that nothing is selected. The complete exploded view should look similar to that shown in Figure 9.31, although the rivet could be on the opposite side, depending on how your pizza cutter was assembled.
6. Click the ConfigurationManager tab (the rightmost tab), below the FeatureManager design tree near the bottom of the window. Then, click the plus sign next to **Default** in the ConfigurationManager. To collapse the assembly, *right click* on **ExplView1** and select **Collapse**. If you want to change any parameters of the exploded view, you can *right click* **ExplView1** and select **Edit Definition**, which will activate the **Assembly Exploder** dialog box.

Congratulations! The pizza cutter assembly is now complete. Save the assembly and close the window.

PROFESSIONAL SUCCESS: THE CONFIGURATIONMANAGER

The ConfigurationManager allows the user to create different configurations of a part or assembly. For example, the guard could have had two configurations: "guard with circular hole" and "guard with rectangular hole". The rest of the guard's features would be the same for both configurations. Using configurations is an easy way to switch between multiple versions of the same part. Using the ConfigurationManager also has advantages within assemblies. In a large assembly, some components may need to be suppressed or hidden in order to allow the assembly to rebuild faster. Different configurations of the same assembly can have different components shown or hidden. For example, one configuration of the pizza cutter assembly could have the guard hidden. Another configuration could show all of the components. Switching between the two configurations would either show the guard or hide it.

The ConfigurationManager is located in the same pane as the FeatureManager design tree. Click the ConfigurationManager tab in the lower left corner of the screen to activate it. To create a new configuration, *right click* **<file name> Configurations**, where **<file name>** is the name of the file that you are working on. Select **Add Configuration** and name the new configuration. Features added to the new configuration are reflected only in that configuration. *Double click* on the original configuration in the ConfigurationManager to switch back to that configuration.

KEY TERMS

Assembly	Configuration Manager	Parallel mate
Coincident mate	Exploded view	Rebuild
Concentric mate	Interference	Selection filter

Assembly Problems

1. Assemble the pizza cutter, beginning with the blade and working toward the handle. Do not use a sub-assembly.
2. Assemble the pizza cutter, beginning with the guard and working toward the handle and then toward the blade.
3. Assemble the handle, cap, and guard into a sub-assembly. Then, assemble the handle sub-assembly to the cutter sub-assembly.
4. Starting with the completed pizza cutter assembly, modify the hole in the cap to be rectangular instead of circular. This will require removing the hole from the original sketch of the cap. Then, create a new sketch plane in the plane of the flat surface of the cap. Make a rectangular cut like the one in the guard. When assembling the revised cap, a new assembly constraint must be added in order to rotate the cap to properly line up with the arms and the rectangular hole in the handle. One possible constraint is to make the long side of the cap's rectangular cut parallel to the long side of the rectangular cut. It may help to hide the guard and cutter sub-assembly. Do an **Interference Detection** check to be sure that the revised cap fits in properly with the rest of the assembly.
5. Model a wooden pencil. Most pencils have three parts: a hexagonal wooden portion with embedded lead (extruded hexagon with a cut at one end to form the point and a cut at the opposite end to fit the cylindrical eraser holder), a metal eraser holder (thin revolve), and the cylindrical eraser. Assemble these three parts to model a pencil.
6. Model a plastic butter tub and cover. Model the tub as a thin revolve. Model the cover as a thick revolve so that the lip at the bottom edge of the cover can be included.

10

Creating Working Drawings

OVERVIEW

Working drawings are the traditional means of graphically displaying a part or assembly. These drawings must provide all of the information necessary to manufacture a part or assembly. Drawings in SolidWorks are based on parts and assemblies that have already been modeled. Views of the models are placed in the drawing and the dimensioned. Notes and a title block are added to the drawing to include pertinent information. Section views and detail views are also used in drawings to better describe the part or assembly.

10.1 DETAIL DRAWINGS IN SOLIDWORKS

Now that you have learned how to model parts and assemblies, you must learn how to put this information into an engineering drawing, or working drawing, the standard engineering graphics format. The engineering drawing presents the details of the part or assembly in a formal manner. There are many variations on the format of engineering drawings. However, one key guideline is that *a drawing should clearly convey all of the information necessary to fabricate the part*. In this chapter, you will learn the common SolidWorks commands for generating engineering drawings.

SECTIONS

- 10.1 Detail Drawings in Solidworks
- 10.2 Editing a Drawing Sheet Format
- 10.3 Creating a Drawing of the Arm
- 10.4 Creating a Drawing of The Pizza Cutter Assembly

OBJECTIVES

After working through this chapter, you should be able to

- Create a new drawing
- Modify a drawing sheet format
- Place and arrange views on a drawing
- Add dimensions to a drawing
- Add tolerances to a drawing
- Display an exploded view in a drawing
- Add a section view to a drawing
- Add a detail view to a drawing
- Label parts and add leader lines to a drawing
- Create a bill of materials

186 Chapter 10 Creating Working Drawings

Prior to solids modeling software like SolidWorks, engineering drawings were laboriously created by draftsmen and designers translating the three-dimensional visual image, either from their minds or a physical model into detailed orthographic projections. The process was slow and changes were difficult. Using SolidWorks, the designer or engineer first creates a three-dimensional virtual model of a part. Based on this model, the SolidWorks drawing mode quickly extracts the two-dimensional orthographic information. Any changes to the three-dimensional model are automatically transferred to the drawing, because of the associative nature of SolidWorks.

In this chapter, you will create drawings of the arm and the pizza cutter assembly. The drawing in Figure 10.1 shows three two-dimensional orthographic views and one three-dimensional isometric view of the arm. Details, such as dimensions, tolerances, and material specifications, are also displayed on the drawing. The title block, which contains drawing control information, is in the lower right corner of the drawing.

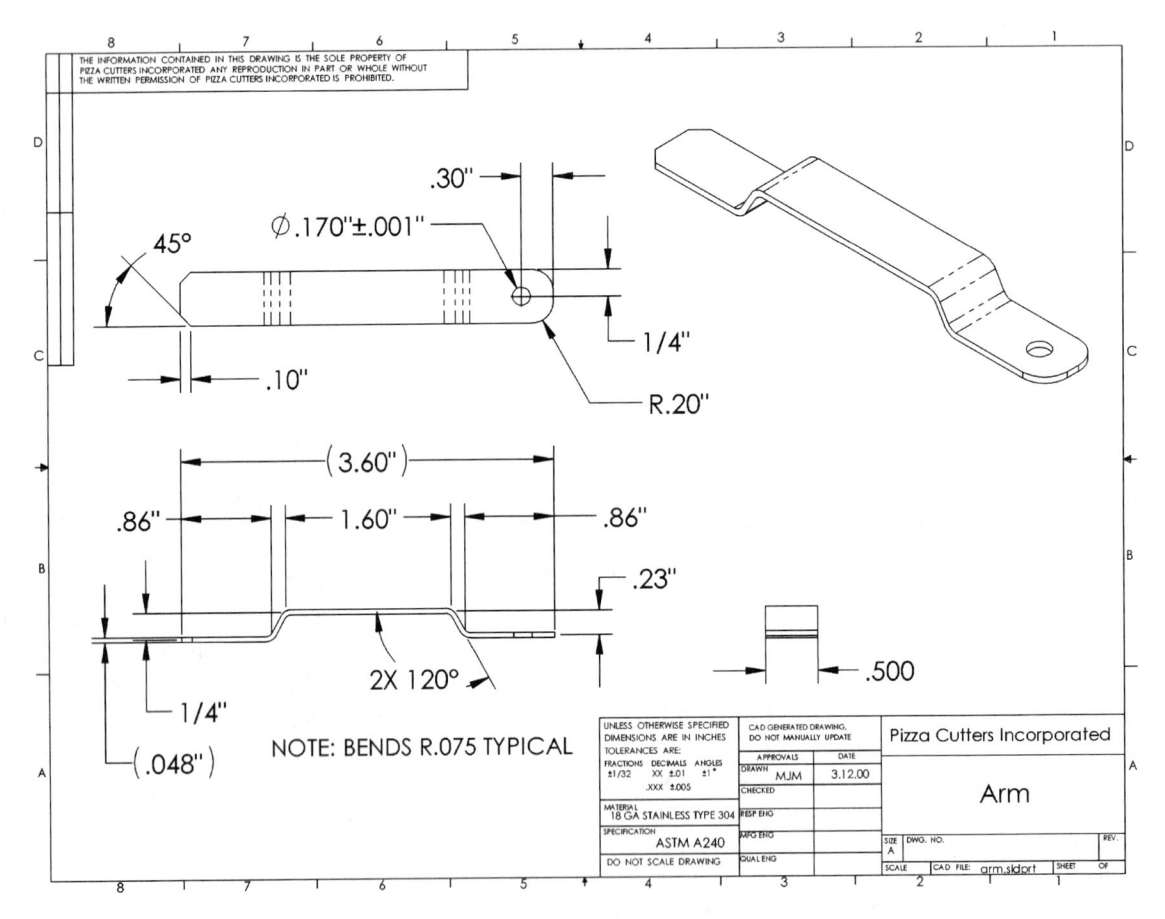

Figure 10.1. Finished arm drawing

10.2 EDITING A DRAWING SHEET FORMAT

In this section, a drawing sheet format that can be used in future drawings will be created. The format can be thought of as the template for all of the drawings that will be created. It includes a border for the drawing, a title block, and a statement that the information is proprietary. The completed drawing sheet format is shown in Figure 10.2.

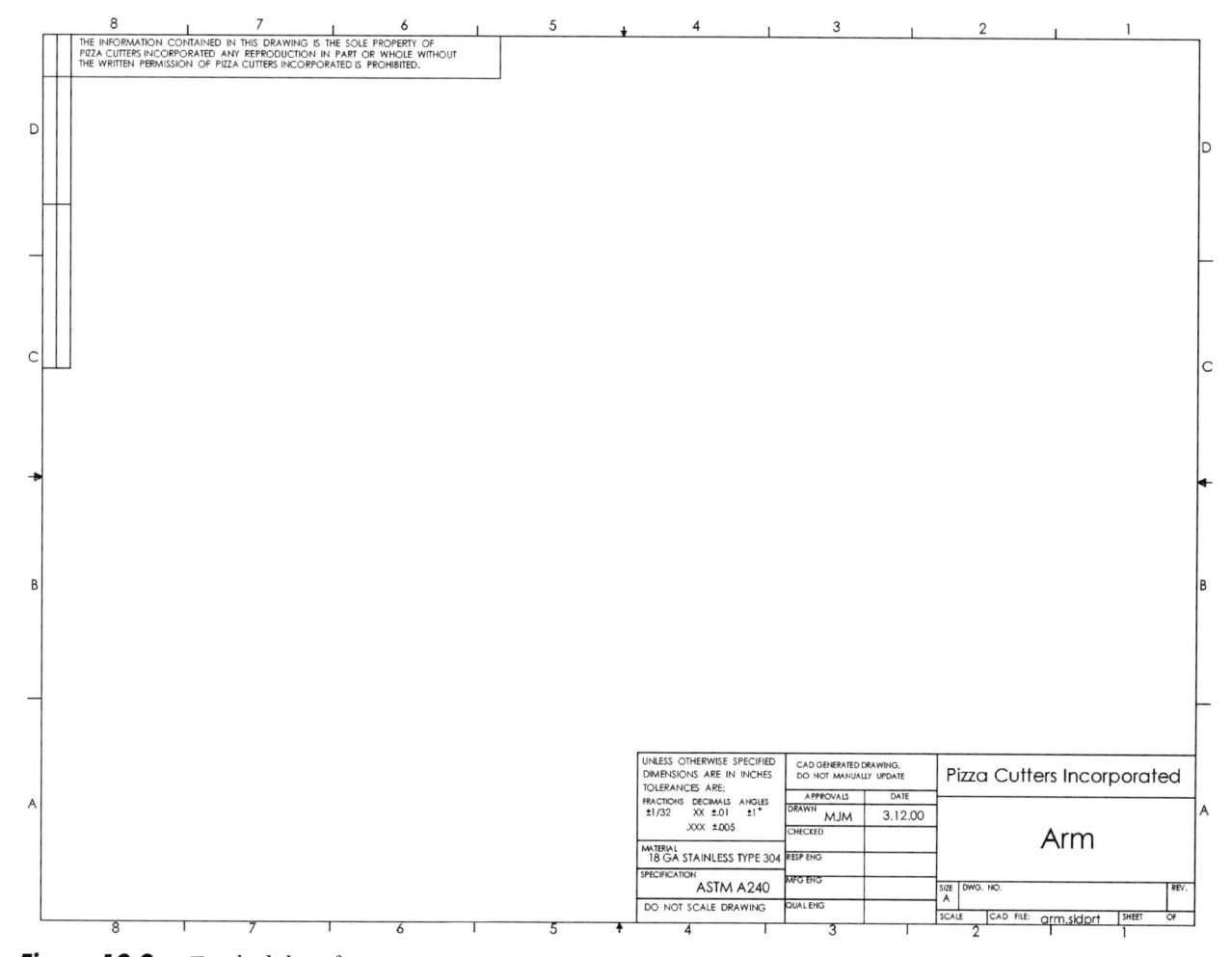

Figure 10.2. Finished sheet format

10.2.1 Creating a New Drawing Document

Creating a drawing in SolidWorks begins by opening a drawing document, which is different from a part or assembly document. In addition to specifying the file type, it is necessary to specify the drawing size. Several standard drawing sizes are available, as indicated in Table 10-1. In this case, an *A* size drawing will be used for the arm of the pizza cutter.

TABLE 10-1 Standard drawing sizes

ANSI (in inches)	ISO (in millimeters)
A 8.50 × 11.00	A4 210 × 297
B 11.00 × 17.00	A3 297 × 420
C 17.00 × 22.00	A2 420 × 594
D 22.00 × 34.00	A1 594 × 841
E 34.00 × 44.00	A0 841 × 1189

1. To open a new drawing file for the arm, click the **New** button in the **Standard** toolbar, or **File ⇒ New**.

2. Choose **Drawing** from the dialog box and click **OK**.

3. In the **Sheet Format To Use** dialog box, select **Standard Sheet Format** with the menu set to **A - Landscape**. The drawing size depends upon the printer you use and what information you must present. Smaller drawing sizes work well for simple parts, whereas larger drawing sizes are necessary to show enough detail for more complex parts. **Landscape** indicates that the longer side of the drawing is horizontal. Click **OK**. A new drawing using the Solid-Works standard sheet format opens, as shown in Figure 10.3.

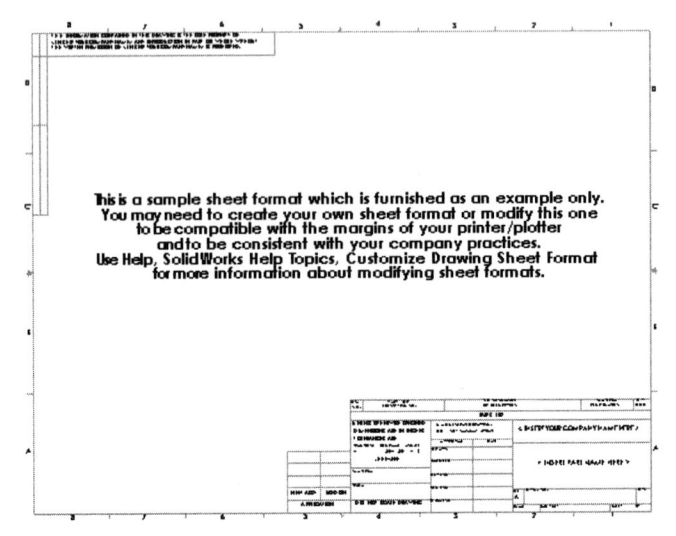

Figure 10.3. Standard sheet format

4. Notice that the icon which looks like a drawing appears next to the cursor when the drawing is opened. This means that the cursor is above the sheet of the drawing. As you will see later, the icon will change when the cursor is over a particular view or other entity (such as a dimension) of the drawing.

5. In the **View** menu, uncheck **Origins**. This removes the origins (that were used in the creation of the parts) from the drawings.

6. In the **View** ⇒ **Toolbars** menu, be sure that **Standard**, **View**, **Annotation**, **Drawing**, **Features**, **Sketch**, **Sketch Relations**, and **Sketch Tools** are checked.

10.2.2 Checking the Options Settings

Before modifying the drawing sheet format, be sure that the SolidWorks settings match the following:

1. Select **Tools** ⇒ **Options** and then select the **System Options** tab followed by the **Default Edge Display** item. In the **Default displays of edges in new drawing views** section, select **Hidden in gray**. This causes hidden lines in the drawing to be displayed as dashed lines in the printed output. The remaining settings should match the ones shown in Figure 10.4.

2. Click the **Document Properties** tab followed by the **Units** item. Be sure that the options are the same as those used to create the parts. Use **Inches**, **Fractions**, 3 for **Decimal places**, 32 for **Denominator**, and **0** for **Decimal places** of angles.

Section 10.2 Editing a Drawing Sheet Format   **189**

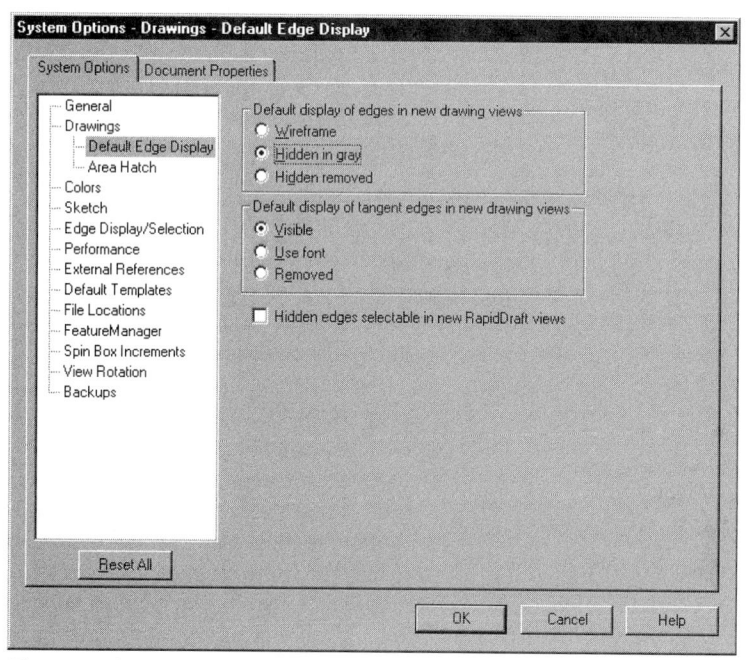

Figure 10.4.   Default Edge Display

3. Select the ***Notes*** item, and click on the ***Font*** button. In this window, both the font type and font size for the notes on the drawing are specified. In the ***Height*** section, click on the ***Points*** button, and then set the size to ***8*** as shown in Figure 10.5. This causes the font size to be eight points. Be sure that the font is set to **Century Gothic** to match this tutorial. Click ***OK***.

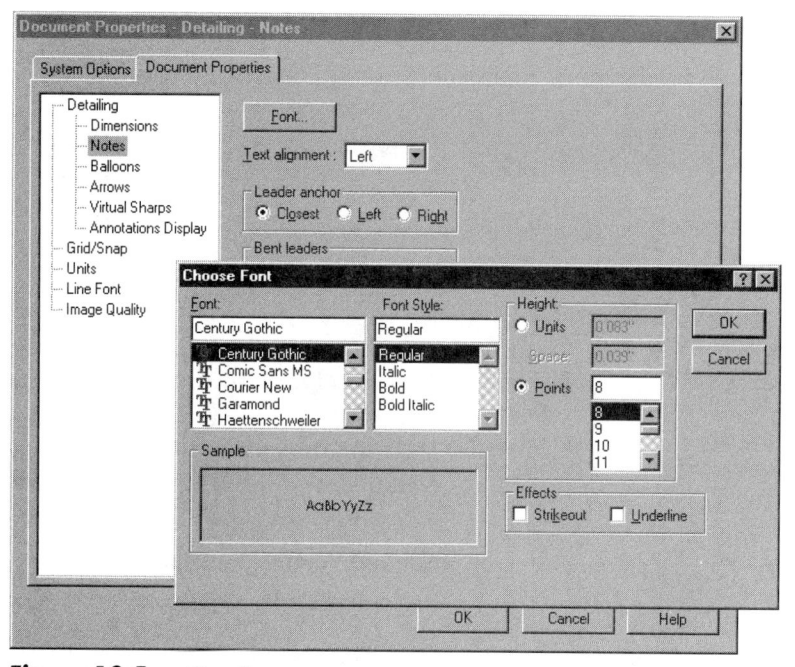

Figure 10.5.   Note font

4. Select the **Dimensions** item, and click on the **Font** button to set the font size for dimensions in the drawing. In the **Height** section, click the **Units** button if it is not already activated. Then, type **3/16** inches as the height. Be sure that the font is set to **Century Gothic** to match this tutorial. Click **OK**.
5. Click **OK** at the bottom of the dialog box in order to apply all of the settings.

10.2.3 The Drawing Environment

The drawing environment in SolidWorks is very similar to the part and assembly environments. Both the FeatureManager design tree and the tools in the **View** toolbar are still active. This section describes the elements and tools used to create drawings from part or assembly files in SolidWorks.

1. The drawing consists of two layers: a format and a sheet. Think of the sheet as a transparent page lying over the format. The format contains items that appear in every drawing, such as the title block, general tolerances, a border, and the scale of the drawing. The sheet contains the drawings, notes, and details specific to a particular part. One drawing can have several sheets, all using the same format.
2. The **Drawing** and **Annotations** toolbars, shown in Figure 10.6, contain the tools both for inserting views of parts onto the sheet and for adding details to the drawing. The toolbars may be vertical, as shown in Figure 10.6, or horizontal. A toolbar can be moved to a different position by clicking on the parallel lines at one end of the toolbar and dragging it to a new location. Hold the cursor over the buttons in the toolbars in order to display the names of the tools. Each of the tools is available from either the **Insert** ⇒ **Annotation** menu, the **Insert** ⇒ **Drawing View** menu, or the **Tools** menu.

Figure 10.6. Drawing and annotations toolbars

The **Drawing** toolbar contains the following tools:
- **Detail View** displays an enlarged portion of an area.
- **Section View** creates a cross-sectional view of a part.
- **Aligned Section View** inserts an aligned cross section.
- **Projected View** creates a new projected view from an existing view.
- **Standard 3 View** creates standard top, front, and right views of a part.
- **Auxiliary View** creates a view of an inclined plane.

- **Named View** creates a chosen view (for example, an isometric view) of a part.
- **Relative View** creates a view based on a user-defined orientation.
- **Align Parallel/Concentric** aligns selected dimensions to be a parallel distance from each other.
- **Crop View** crops the selected view.

The **Annotation** toolbar contains the following tools:

- **Note** adds a textual description to a selected portion of the drawing.
- **Surface Finish** displays the symbol used to specify the surface finish of a part.
- **Geometric Tolerance** displays the symbols that describe tolerances on both dimensions and constraints.
- **Balloon** attaches a balloon note to the selected edge.
- **Datum Feature Symbol** attaches a symbol that indicated a datum plane to a selected edge or surface.
- **Weld Symbol** inserts a symbol that designates a weld.
- **Datum Target** inserts a datum target symbol.
- **Block** creates or inserts custom symbols for standard drawing items.
- **Model Items** imports dimensions, annotations, and reference geometry from the part or assembly.
- **Center Mark** creates a center mark for a selected circle or arc.
- **Hole Callout** inserts text related to a hole.
- **Cosmetic Thread** adds a simplified schematic thread to a cylindrical feature.
- **Stacked Balloon** creates a series of balloons with a single leader.

10.2.4 Modifying the Format's Text

The standard SolidWorks format can be customized to suit the project or company needs. In this section, the standard format shown in Figure 10.3 will be modified into custom template, shown in Figure 10.2.

1. With the drawing symbol next to the cursor, *right click* on the sheet and select **Edit Sheet Format**. Or, in the **Edit** menu select **Sheet Format**. The text, the boxes, and the text within the boxes can now be edited.
2. Click on the large block of text in the middle of the drawing. Four green squares appear at the corners of the text, signifying that it is selected. Press delete on the keyboard to remove the text.
3. Use the **Zoom To Area** tool to zoom to the bottom right corner of the drawing in order to view the title block.
4. With the **Select** tool, *right click* <INSERT PART NAME HERE> and then click **Properties**, or select it and use **Edit ⇒ Properties**. A *Properties* window appears, as shown in Figure 10.7. Change the <INSERT PART NAME HERE> to **Arm**. While still in the dialog box, be sure *Use Document's Font* is *not* selected, because the font that was specified in the **Options** dialog box is too small for this item. Click the **Font** button. In the window that appears, the *Font*, the *Font Style* (e.g., Plain or Bold), and the *Height* can be changed. Set the *Height* of this text to be *18 points*. Click **OK** twice.

192 Chapter 10 Creating Working Drawings

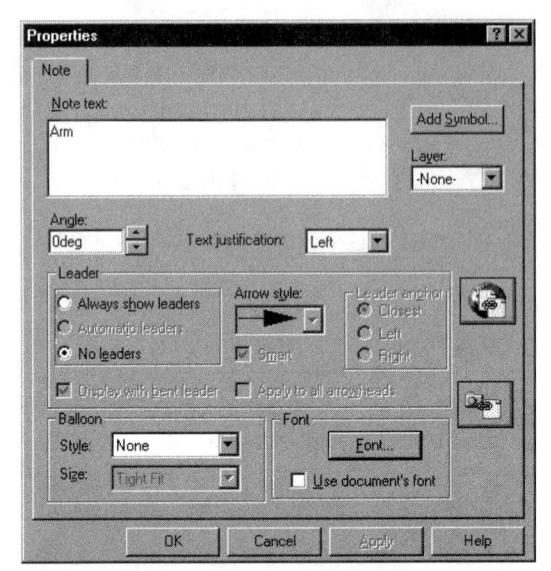

Figure 10.7. Properties dialog box

5. To move the text to the center of the box, click on the word **Arm** to select it, and then drag it to the center of the box. It might help to click the **Redraw** button in the **Standard** toolbar to clean things up.

6. In the same manner, change **<INSERT YOUR COMPANY NAME HERE>** to an appropriate name, and change the font size to 12 point. Center the text in the box.

7. On the left side of the title block, change **FINISH** to **SPECIFICATION**.

8. In the **MATERIAL** field, change the dash to **18 GA STAINLESS TYPE 304**, modify the font size to seven points by typing in **7**, and center the text in the box. "**GA**" refers to the gauge, or thickness, of the sheet metal. For stainless steel sheet metal, 18 gauge is .048 inches thick.

9. In the **SPECIFICATION** field, change the dash to **ASTM A240**, a stainless steel specification, and center the text in the box.

10. Click anywhere in an open space on the sheet format. A small cross appears where you clicked. Click the **Note** button in the **Annotations** toolbar, or select **Insert** ⇒ **Annotations** ⇒ **Note**. The *Properties* dialog box appears. In the *Note Text* field, type in *arm.sldprt*. Click **OK**. Move the text so that it is next to the text **CAD FILE:**, below the text **Arm**. It may help to zoom in to place the text.

11. With the **Select** tool, click the **NEXT ASSY** line of text, found on the left side of the title block. Remove the text using the delete key. *Control click* all of the text in the drawing that is not shown in Figure 10.8. Click on five or six items at a time. Then, delete the text. Be sure that you do not delete the sheet format. This may occur if you accidentally click on an area of the drawing that is not text. If this happens, the item listed in the *Confirm Delete* dialog box will be *Sheet Format1 (Drawing Sheet)*. Click *No* to cancel the delete. If you do accidentally delete the sheet format, click the **Undo** toolbar button to recover it. When you are finished removing the text, the screen should look similar to that shown in Figure 10.8.

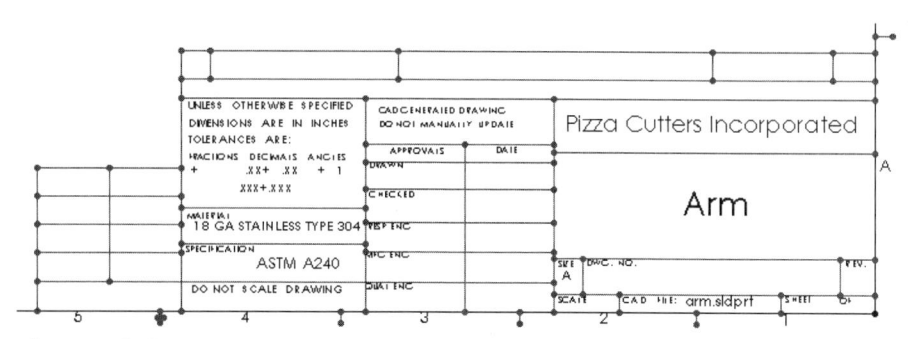

Figure 10.8. Text placed in title block

12. **Zoom To Fit** and then **Zoom To Area** in the upper left corner of the drawing. With the **Select** tool, *right click* the second line of text and click **Properties**. In the **Properties** dialog box, change **<INSERT YOUR COMPANY NAME>** to an appropriate name. Do the same for the third line of text.

10.2.5 Modifying the Sheet Format's Lines

Now, extra lines in the title block will be removed. When modifying the lines, keep in mind that the format can be used for other parts, so there will be some blank boxes.

1. Be sure that you are still editing the format (and not the sheet) by clicking the **Edit** menu. If **Sheet Format** is grayed out, then you are still editing the format. Do *not* click on **Sheet**. This will toggle the edit to the sheet.

2. Zoom to the title block in the bottom right corner of the drawing. *Control click* and delete the six lines in the title block, as shown in Figure 10.9. Selected lines are highlighted in green. The line that is too long will be shortened in the next step. If you accidentally select the sheet format, the **Confirm Delete** dialog box will appear. Click **No** to avoid deleting the format.

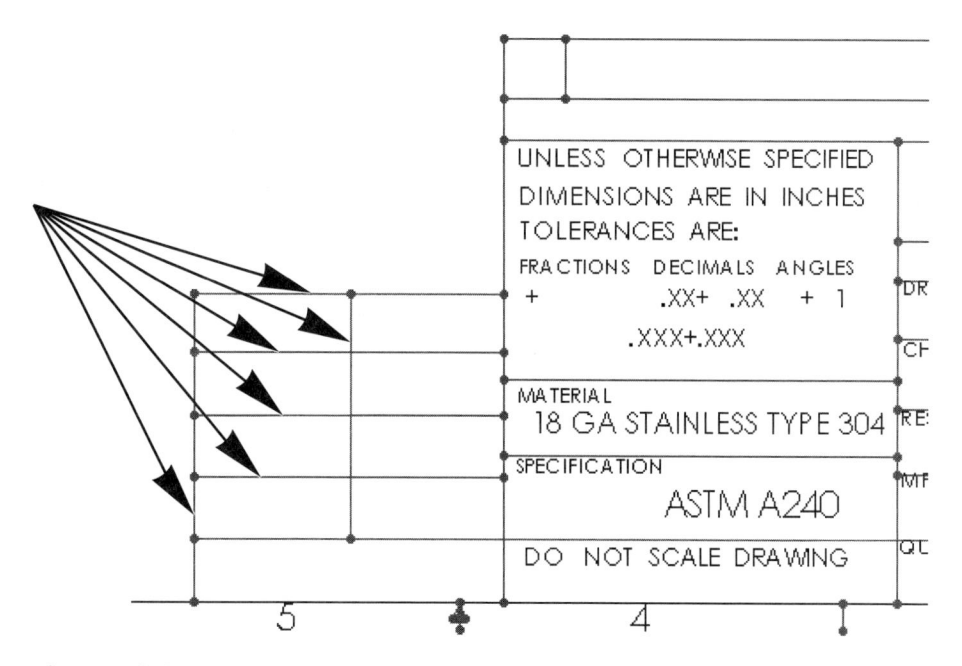

Figure 10.9. Lines to delete

3. To change the length of a line, select the line with the **Select** tool. Then, drag the end of the line to the desired length. Shorten the line that extends past the block so that it meets the leftmost vertical line. A lightbulb appears next to the cursor when the line's endpoint is over a vertical line.

4. Remove the lines at the top of the title block, leaving only those shown in Figure 10.10.

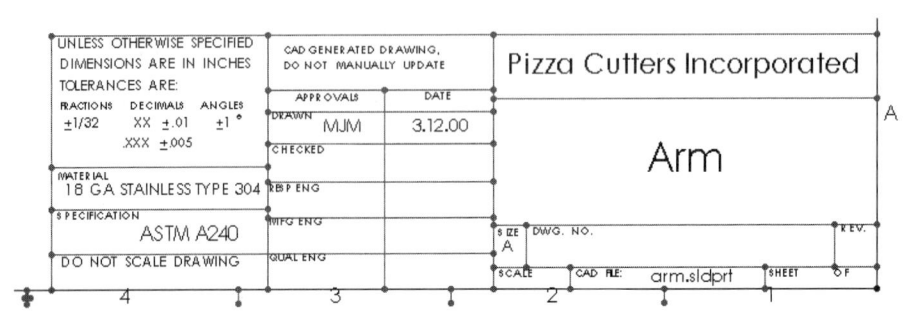

Figure 10.10. Finished title block

5. Figure 10.10 indicates the values of the drawing's general tolerances. A dimension that has two digits after the decimal point indicates that the tolerance is plus or minus .01 inches. In other words, the dimension of the actual part can be .01 inches larger or .01 inches smaller than the specified dimension and still be acceptable. Update the tolerance note to reflect the tolerances of the drawing, as shown in Figure 10.10. You can find both the ± symbol and the degree symbol by clicking the **Add Symbol** button in the **Properties** dialog box in order to activate the **Symbols** dialog box, as shown in Figure 10.11. These symbols are in the **Modifying Symbols** symbol library. Be sure that the **Use Symbol** button is on. Adjust the spacing as needed, so that the modified text is under the correct heading in the title block.

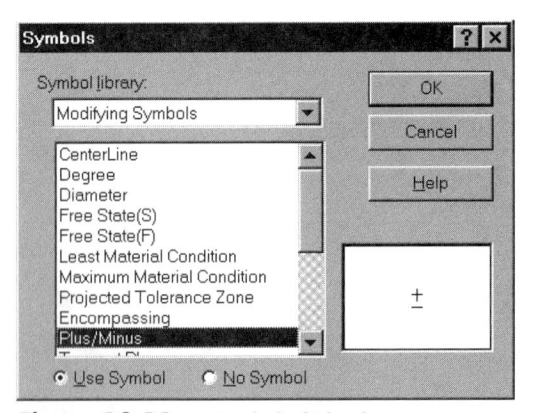

Figure 10.11. Symbols dialog box

6. Click on an open space on the screen. Using the **Note** tool, add your initials in the **DRAWN** box of the format. Be sure the *No leaders* button is selected in the **Properties** dialog box. Repeat for the date.

7. Lines and other entities can be created in sheet formats using the same sketch tools used in creating parts. For example, more boxes could be created in the title block by using the **Line** tool. Figure 10.10 shows the completed title block.
8. *Right click* on any open space on the sheet format and select **Properties**. The **Sheet Setup** dialog box appears. In the **Type of projection** section select **Third angle**, which is the standard projection system in North America. Click **OK**.

Once you are satisfied with the sheet format, save it for future use. Select **File ⇒ Save Sheet Format** and click the **Custom Sheet Format** button in the dialog box. Click **Browse** to locate the folder or device in which you want to save the format. Save it as *tutorial format*. This format can be used for any new drawing.

10.3 CREATING A DRAWING OF THE ARM

In this section, several views of the arm will be inserted into the drawing that is currently open. You will also add the dimensions of the arm in order to fully describe its geometry. The finished drawing of the arm is shown in Figure 10.12.

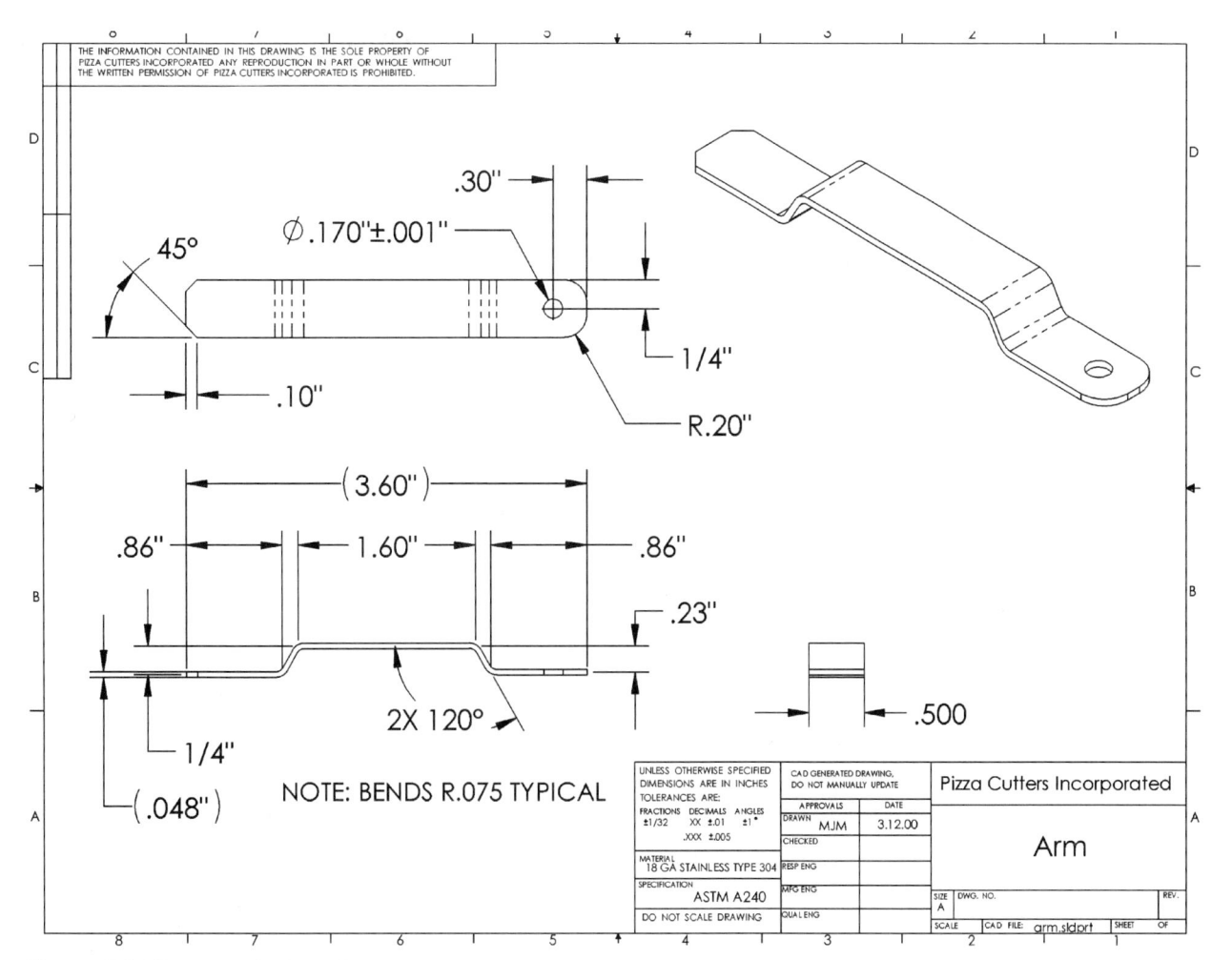

Figure 10.12. Finished arm drawing

10.3.1 Placing Orthographic Views

Three orthographic views are customarily used in engineering practice to describe the geometry of a part. These views, the top, front, and right views, can be placed into a SolidWorks drawing directly from a part with just a few clicks of the mouse. Be sure that the drawing sheet format that you created in the previous section is still open.

1. To begin, open the file of the part depicted in the drawing. Click **File** ⇒ **Open**, and open the arm part file that you saved in an earlier chapter. Be sure that the *Files of type:* is set to *Part Files* or *SolidWorks Files* in the **Open** dialog box. A new window opens that contains the model of the arm. Return to the drawing window by selecting **Window** ⇒ **Draw1-Sheet1**. **Zoom to Fit** to show the entire drawing.

2. Select **Edit** ⇒ **Sheet**, so that you will edit the sheet and not the sheet format. The format (border and title block) should turn from blue to gray, indicating that it cannot be modified.

3. Click on the **Standard 3 View** button (three views) in the **Drawing** toolbar, or select **Insert** ⇒ **Drawing View** ⇒ **Standard 3 View**. Notice that the cursor includes a three-dimensional box, indicating that a part should be selected.

4. Activate the **arm** window by selecting **Window** ⇒ **arm**. The cursor should still have a box next to it.

5. Click anywhere on the arm to select it. Upon selecting the arm, SolidWorks automatically returns to the drawing window with the top, front, and right views placed on the sheet, as shown in Figure 10.13.

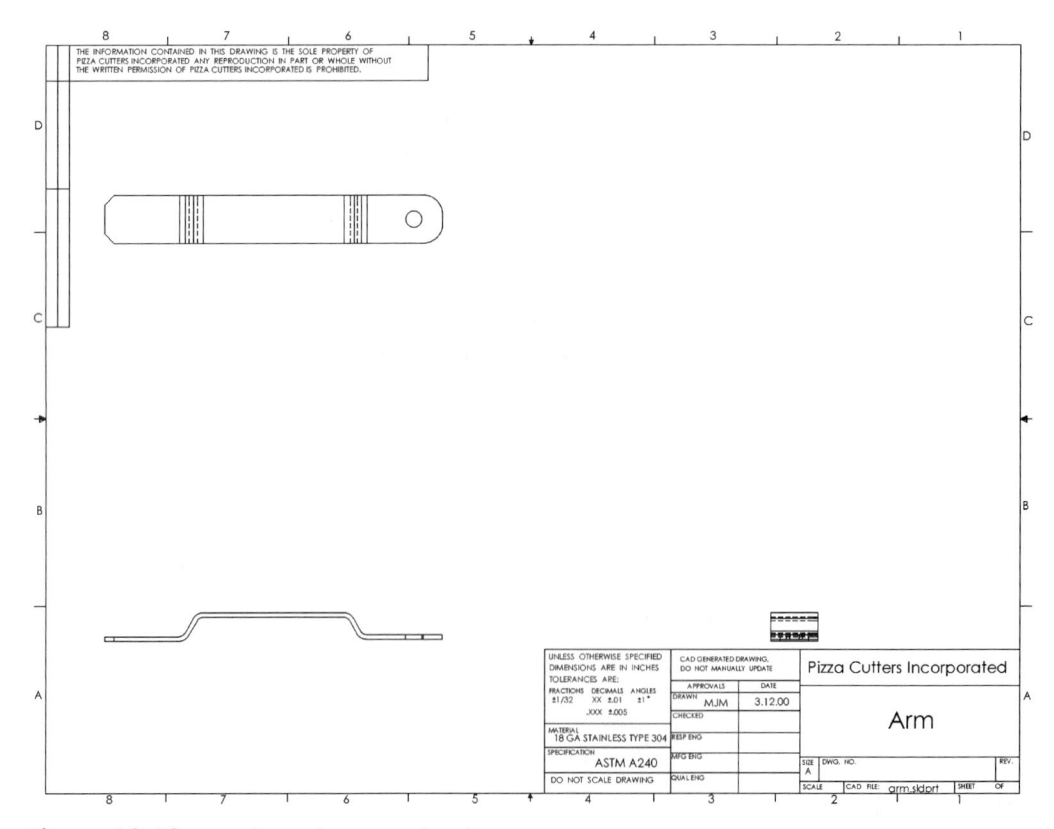

Figure 10.13. Orthographic views placed

6. Bring the cursor over one of the views in the drawing. When the cursor is inside the box that surrounds the view, but is not over the part itself, the cursor icon is a component surrounded by the corners of a box. In this mode, the cursor will select the view of the component. When the cursor is over the features of the component (such as faces, lines, or points), an icon of that feature appears next to the cursor. In this mode, the cursor will select the feature.

10.3.2 Adding a Named View

Named views are preset orientations of the part from the part document. The most commonly used named view is the isometric view. Including an isometric view helps the person reading a drawing to visualize a three-dimensional image of the part.

1. Click the **Named View** button (a view with N) in the **Drawing** toolbar, or select **Insert ⇒ Drawing View ⇒ Named View**. Notice that the cursor has changed to a three-dimensional box, indicating that a part should be selected.
2. Return to the **arm** window and click on the arm. The **Named Model View** dialog box appears. Any view of a part can be placed in a drawing. In this case, you want to place an isometric view, so *double click* **Isometric** from the list, or click **Isometric** and then **OK**.
3. The dialog box closes and the cursor changes to a cross. Select **arm-Sheet1** from the **Window** menu. As the cursor moves across the drawing, the isometric view moves with it. Click on the upper right portion of the drawing to place the isometric view. Answer **Yes** to the dialog box giving the option to use true isometric dimensions rather than projected dimensions. The drawing should look similar to that shown in Figure 10.14.

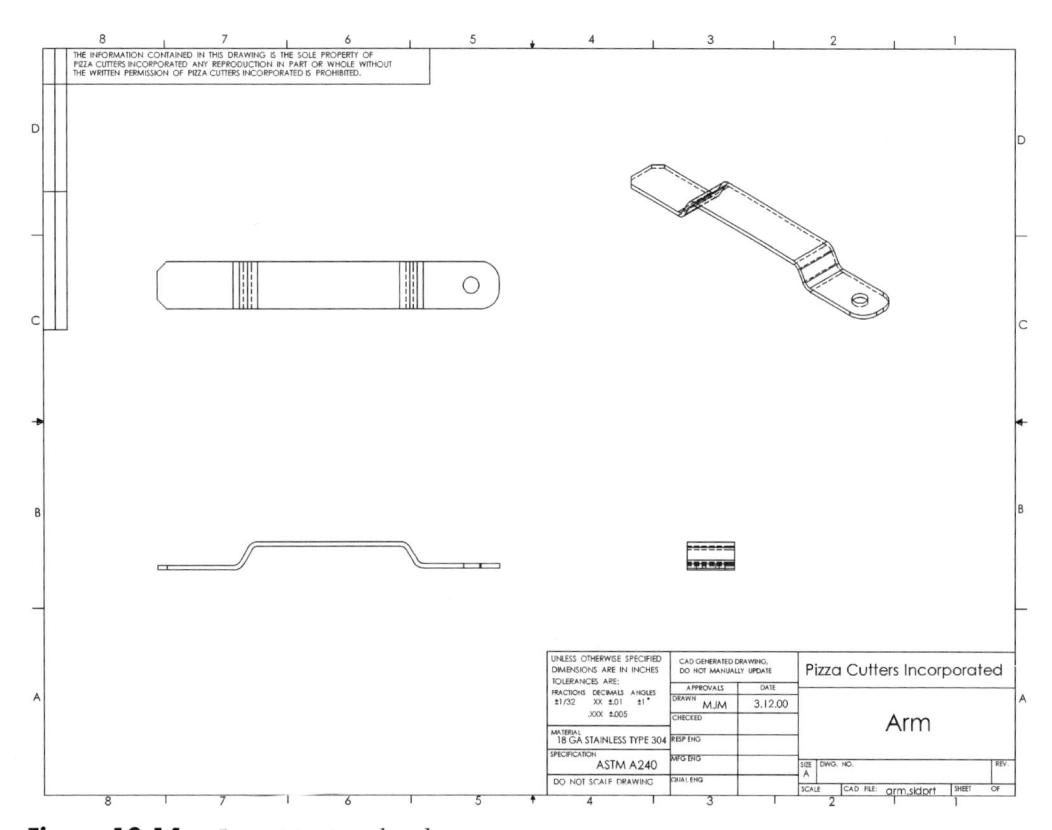

Figure 10.14. Isometric view placed

10.3.3 Adjusting the Views on the Sheet

Views can be moved or modified in several ways. The positions of the views can be changed either to clarify the drawing or to make room for more views on the drawing. The scale of a view can be changed to see more detail. The display for each view can be set to **Wireframe**, **Hidden in Gray**, or **Hidden Lines Removed**, depending upon which display is optimal.

1. To ensure that the views do not overlap the sheet format or other views, the views may have to be moved. Click on the box that surrounds the front view (the bottom left view), but not the arm itself. The box around the view becomes highlighted in green. To move the view, click and drag the green bounding box. Be sure not to click on one of the eight green squares. The icon next to the cursor must appear as perpendicular lines with arrows at the ends in order to move the view. Notice that the other two orthogonal views are forced to move with the front view, since they are projections of this principal view. Move the front view to match Figure 10.14.

2. The right view (lower right) is somewhat difficult to move, because the box enclosing this view is very close to the part. Zoom in on the right view. Select the box enclosing it, so that it is highlighted in green. Then, position the cursor over a corner of the box so that a diagonal arrow appears next to the cursor. Drag the cursor to make the box larger. Now it should be easier to move the larger box after you **Zoom To Fit**. Move the views to match Figure 10.14. Notice that both the right and top views are constrained to stay aligned with the front view. You can preview the printed drawing at any time by clicking the **Print Preview** button in the **Standard** toolbar or by selecting **File** ⇒ **Print Preview**. Click **Close** to return to the drawing.

3. Click within the box that surrounds the isometric view (but not on the arm itself) to select this view. Zoom in on the isometric view of the arm. The hidden lines are somewhat confusing in this view. Click the **Hidden Lines Removed** button in the **View** toolbar to remove the hidden lines. Repeat this for both the top view (upper left) and right view (lower right).

4. The current display of the drawing shows tangent edges between a curved surface and a flat surface as a solid line. This occurs at the bends and the filleted corners of the arm. There are three ways to display tangent edges: **Tangent Edges Visible** (solid line), **Tangent Edges With Font** (dashed line), **Tangent Edges Removed** (no line). To change the display of tangent edges for the isometric view, *right click* on the view, and then select **Tangent Edge**. Click on **Tangent Edges With Font** in this case. The isometric view should look like Figure 10.15. Repeat this for the top view. For the front and right view use **Tangent Edges Removed** for clarity.

5. Each view within a drawing can have its own scale. To make the isometric view larger, select the isometric view and click **Edit** ⇒ **Properties**. In the ***Drawing View Properties*** dialog box that appears, deselect ***Use sheet's scale***. Set the scale to **3:2**, as shown in Figure 10.16. Click **OK**. The isometric view enlarges to be 1.5 times larger than the other views. The drawing should look like Figure 10.15.

Section 10.3 Creating a Drawing of the Arm   **199**

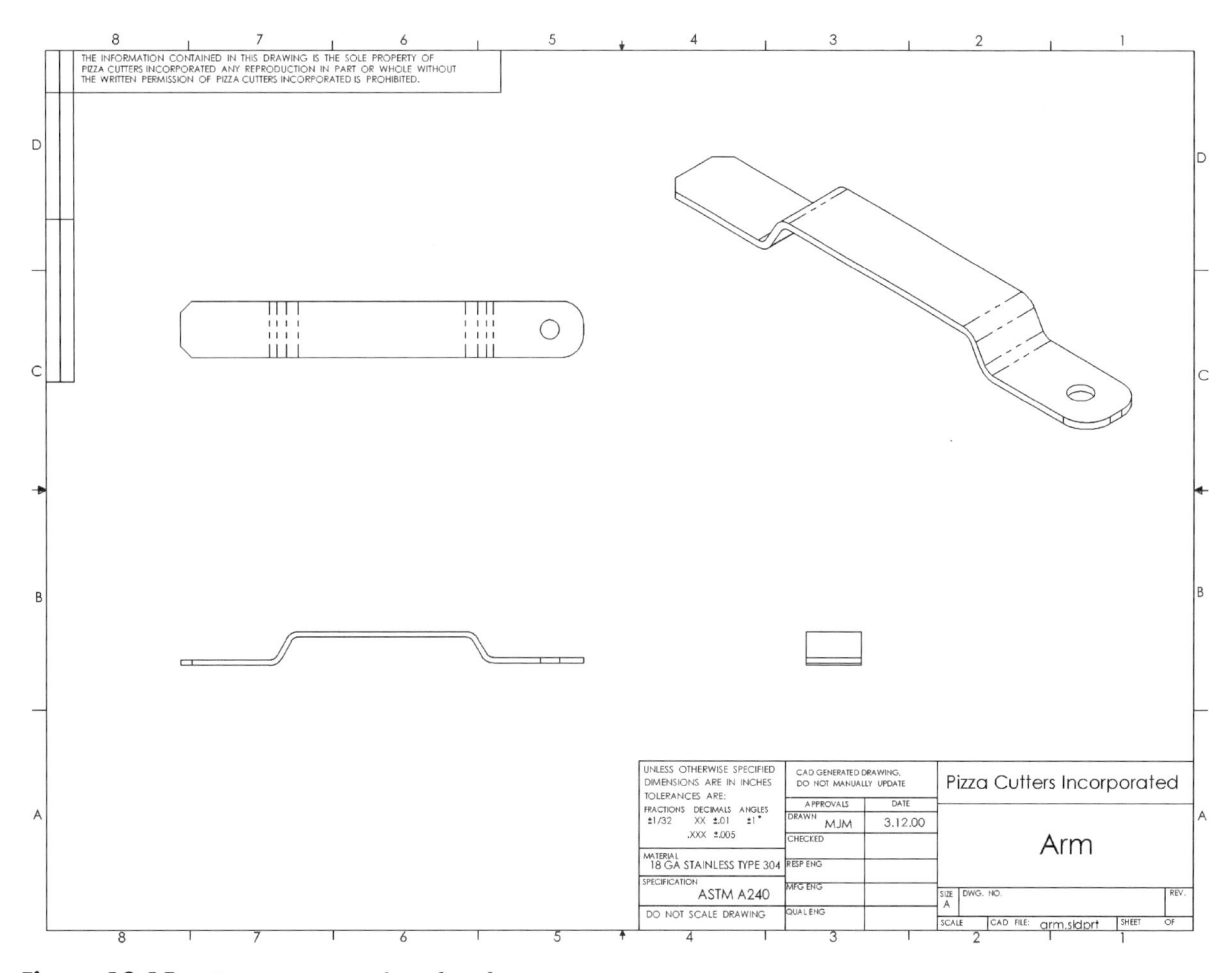

Figure 10.15.   Isometric view enlarged on drawing

Figure 10.16.   Drawing View dialog box

10.3.4 Adding Dimensions to the Orthographic Views

Dimensions that were used during the creation of the part may automatically be placed on the views of the drawing. Dimensions placed in the drawing are the current dimensions of the part. If the part geometry is changed later, the drawing will automatically update to reflect the changes. Likewise, if the dimensions on the drawing are changed, the part will be updated automatically.

1. Select the top view (upper left), so that it is highlighted in green. Click the **Model Items** button in the **Annotations** toolbar, or Select **Insert ⇒ Model Items**. In the *Insert Model Items* dialog box, as shown in Figure 10.17, be sure that the *Dimensions* box is checked. Click **OK**. The dimensions for the view are displayed. Do not worry if the dimensions are hard to read. They will be cleaned up shortly.

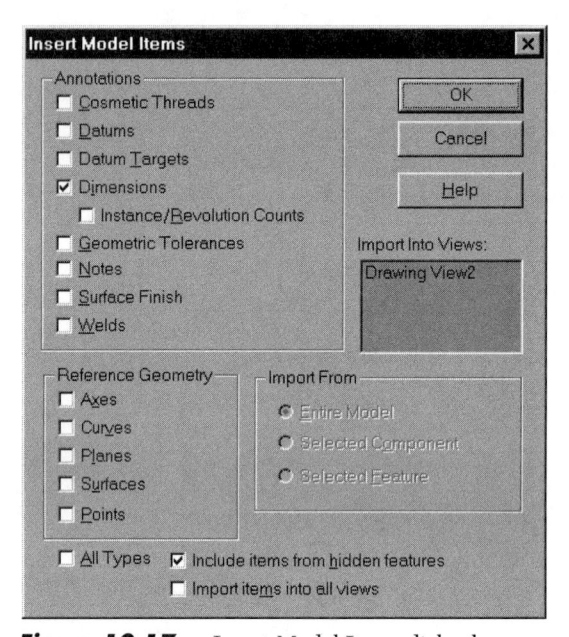

Figure 10.17. Insert Model Items dialog box

2. In a similar manner, place dimensions on the front view. If you try to place dimensions on the right view, no dimensions will appear, because all of the necessary dimensions have already been shown in the other two views. In fact, this view is not even necessary in this drawing. All of the details of the arm are shown in the front and top views. Nevertheless, the right view will be retained for completeness.

3. Clean up the dimensions by dragging them just as you did when creating parts. In placing dimensions, the most important thing to remember is that they should be both clear and in a logical order. A dimension can be moved between views by holding down the shift key on the keyboard and dragging it to another view. It is necessary to drag the dimension all the way onto the new view before releasing the mouse button. Once the dimension is in the new view, drag the dimension to the desired location. Some dimensions cannot be moved to certain views, because the dimension is in a plane perpen-

Section 10.3 Creating a Drawing of the Arm   **201**

dicular to the plane of the view. Move the dimensions to the positions shown in Figure 10.18. The dimensions for the chamfer on the left end of the top view may be placed differently than those shown in the Figure.

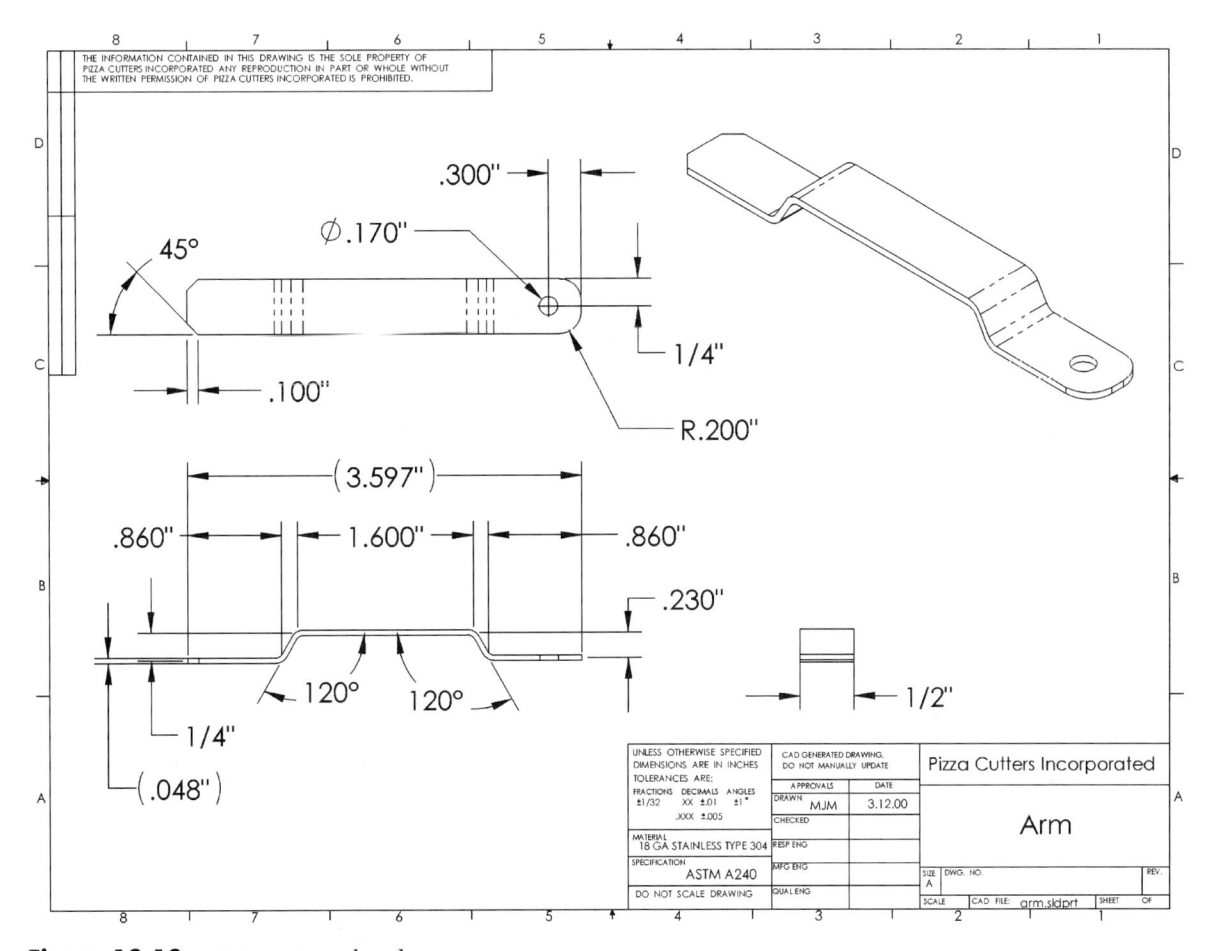

Figure 10.18.   Dimensions placed

4. Some dimensions flip when they are moved. To see this, move the **1/4** dimension on the front view horizontally. The dimension should flip as it passes the left end of the arm. To make this dimension appear as it does in Figure 10.18, move it so that the arrows are positioned as in the Figure. While the dimension is still selected, click the green box where the lower arrow makes a right angle. This flips the position of the **1/4** dimension to that shown in Figure 10.18.

5. The arrowheads on some small dimensions may overlap. To change the arrowhead positions, *right click* on the dimension. Then, click **Properties** to open the ***Dimension Properties*** dialog box. Click **Outside** in the ***Arrows*** field, then **OK**. Another method is to select the dimension and then click the leftmost **Arrow In\Out\Smart** button in the PropertyManager, as shown in Figure 10.19. Do this for the .170, .300, 1/4, and .200 dimensions of the top view, the .048, 1/4, and .230 dimensions of the front view, and the 1/2 dimension of the right view.

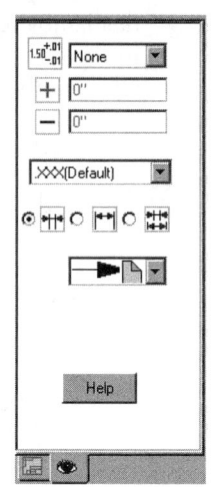

Figure 10.19. Property Manager

6. On a drawing center marks should be added to circles. To do this, click on the **Center Mark** button in the **Annotations** toolbar. The icon next to the cursor turns to a cross. Click on the .170 hole in the top view to add center marks to the hole. It may be necessary to zoom in to see the center marks.

7. Sometimes the dimensions used to model the part are inadequate to clearly represent its dimensions on the drawing. For instance, the overall length of the arm is not evident in the drawing. However, a Reference Dimension can be displayed on the drawing to indicate the overall length. Reference dimensions cannot be used to "drive" the model of the part. In other words, changing a reference dimension on the drawing cannot change any dimension on the part model. To add a reference dimension for the overall length of the arm, activate the **Dimension** tool. In the front view, click on the left end of the arm, followed by the right end of the arm, and place the dimension as usual. The dimension will have parentheses around it, indicating that it is a reference dimension.

8. Because the text "**18 GA**" in the title block defines the arm's thickness to be .048 inches, the .048 dimension in the drawing should be displayed as a reference dimension, with parentheses. To do this, *right click* the **.048** dimension and select **Display Options** ⇒ **Show Parenthesis** in the menu that appears. Parentheses now surround the dimension, indicating that it is a reference dimension. Unlike the previous reference dimension, this dimension "drives" the model of the part. Changing it while change the part.

9. The radius of the bends did not appear when dimensions were automatically placed on the views because the **Auto Fillet** option was used when the arm was extruded. To add a note specifying the bend radius, click in the open space below the front view and then click on the **Note** toolbar button. Type "**NOTE: BENDS R.075 TYPICAL**" in the *Note Text* box. Change the font to 16 point and click **OK** twice. Then adjust the position of the note.

10.3.5 Modifying a Dimension's Text

Text can be added to dimensions in order to better describe a part. One example is when two (or more) dimensions refer to similar features. In this case, one of the dimensions can be removed, and a "**2X**" or a "**2 PLACES**" can be added to the other dimension, indicating that the dimension occurs two times. This saves space and simplifies drawings of parts with several similar features.

1. One of the two **120°** dimensions can be removed if the dimension of the remaining angle indicates that the angle occurs twice. Delete the left dimension by selecting it and the hitting delete key on the keyboard.
2. *Right click* the other **120°** dimension and select **Properties** from the menu that appears.
3. The **Dimension Properties** dialog box appears. Click the **Modify Text** button. Type "**2X**" (without quotes), followed by a space in front of **<DIM>**, as shown in Figure 10.20. The **<DIM>** refers to the numerical value of the dimension. The **Preview** field at the bottom of the dialog box shows what the dimension will look like on the drawing.

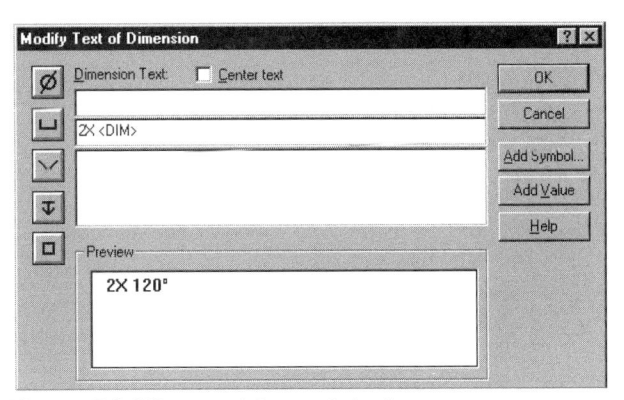

Figure 10.20. Modify Text dialog box

4. Click **OK** twice to return to the drawing. The **120°** dimension is shown in Figure 10.21.

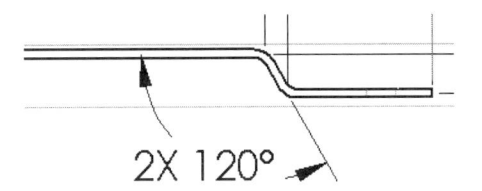

Figure 10.21. 2X dimension added

10.3.6 Specifying Tolerances

The tolerance for a dimension is specified according to the number of digits it has past the decimal point. However, the number of digits of the dimensions must be changed to denote the correct tolerance.

1. The radius of the round at the right end of the arm is not a critical dimension and does not need a tight tolerance. To change the number of digits after the decimal, select the **R.200"** dimension. In the PropertyManager window shown in Figure 10.19 click on the pulldown menu with **.XXX(Default)**. Select **.XX**. The text on the dimension updates to reflect the change.
2. Repeat Step 1 to update the other dimensions with trailing zeros, so that the dimensions look similar to those shown in Figure 10.22.
3. The width of the arm is critical. Its dimension must be accurate so that it fits snugly into the handle. Right now, the dimension is specified as **1/2**, indicating that the tolerance should be ±1/32 inches. To make this tolerance tighter, *right click* on this dimension and select **Properties**. Uncheck

204 Chapter 10 Creating Working Drawings

Use Document's Units and click the **Units** button. In the **Dimension Units** dialog box, click the **Decimal** button. Click **OK** two times. The dimension updates to become a decimal with three places of precision.

4. The tolerance on the .170 hole in the arm should be ±.001. But the general tolerances in the title block only call for a tolerance of ±.005 for a dimension with three digits past the decimal point. To add the tolerance to the dimension select the **.170** dimension. At the top of the PropertyManager window click on **None** in the **Tolerance display** pulldown menu and select **Symmetric**. Type **.001** in the **Maximum variation** box just below. Then click anywhere in the Graphics Window to show the tolerance. The drawing with these changes should look like that shown in Figure 10.22.

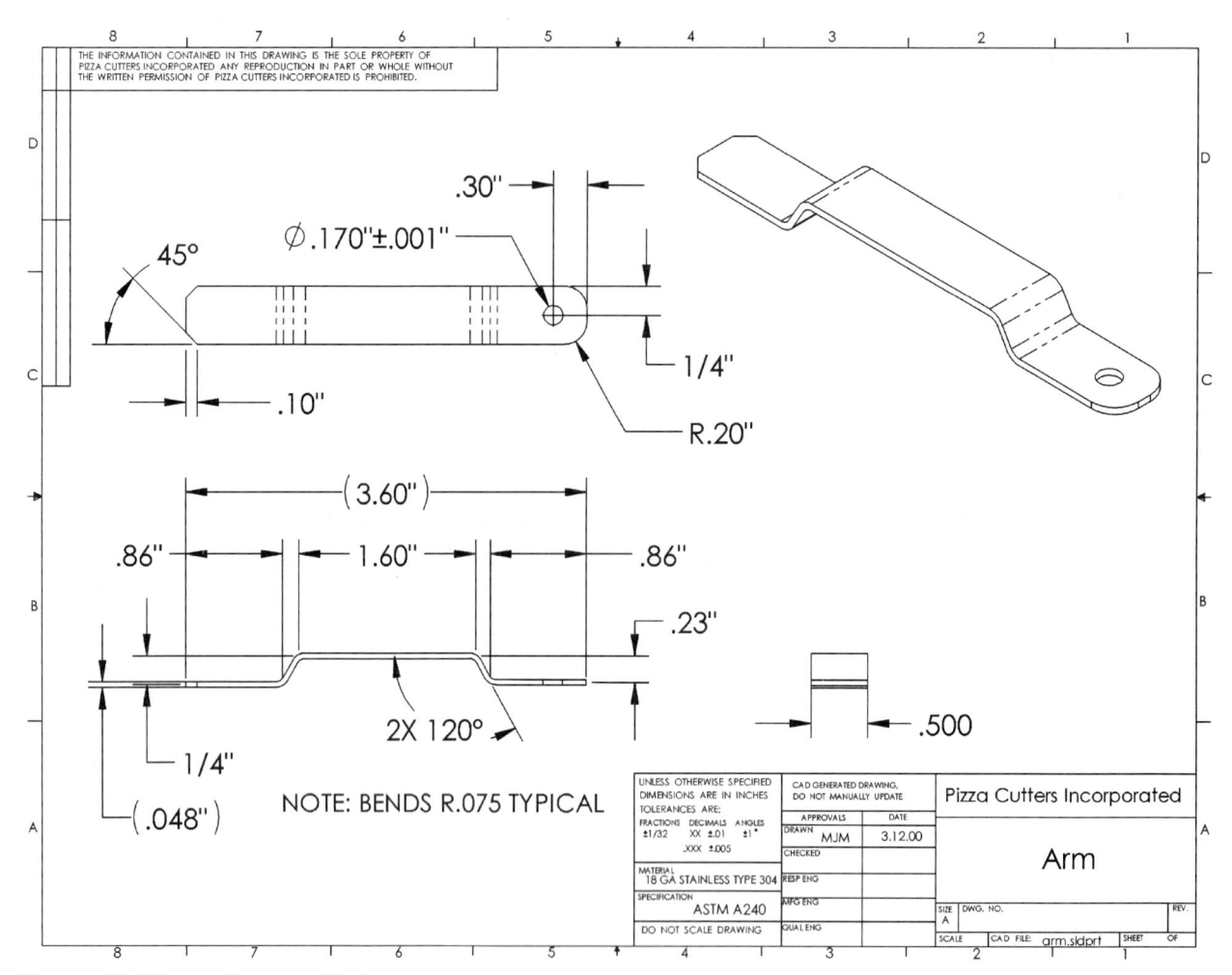

Figure 10.22. Tolerances added

10.3.7 Changing Dimensions in the Drawing and the Part

Dimensions changed in the drawing change the geometry of the solid model, because of the associative nature of SolidWorks. Likewise, dimensions changed in the part are reflected in the drawing. The length of the arm in the drawing will be modified and then changed back in the part, to demonstrate the codependence between the drawing and the model in SolidWorks.

1. *Double click* the **1.60** dimension in the front view of the arm. In the **Modify** dialog box, set the value to **3** inches and click on the green check. The part will not update until the document is rebuilt. Click the **Rebuild** button (a stoplight) in the **Standard** toolbar, or **Edit** ⇒ **Rebuild**, to update the document. All of the views change, including the isometric view. For now, it is acceptable if the drawing extends past the borders of the sheet. You will be changing the arm back to its original length shortly.

2. Select **Window** ⇒ **arm** to activate the **arm** window. The arm reflects the dimension change and is now longer.

3. *Double click* the arm to show all of the dimensions. In the **Orientation** dialog box, set the view to **Front**. *Double click* the **3** dimension and change it back to **1.6**. Click the **Rebuild** button in the **Standard** toolbar.

4. Return to the drawing window. The drawing has changed to reflect the changes made in the part document. The associative character of both the part and the drawing is a powerful feature and should be used carefully. A small change in a drawing document could adversely affect the assembly that contains that part.

Congratulations! This completes the drawing of the arm. The drawing should look like Figure 10.12. Check the preview of the drawing by clicking **Print Preview** in the **Standard** toolbar. Print the drawing by selecting **File** ⇒ **Print** or clicking the **Print** button in the **Standard** toolbar. Save the drawing as *arm drawing* and close both the drawing window and the arm window. Answer **Yes** to the dialog box warning about the referenced models.

PROFESSIONAL SUCCESS: COSMETIC THREADS

Cosmetic threads are a schematic representation of a threaded feature, such as a threaded hole. Instead of displaying the actual threads, a hidden line and a Thread Callout are placed both on the drawing and on the part. The hidden line shows the minor diameter for an external thread (e.g., a bolt) or the major diameter for an internal thread (e.g., a nut). Both internal and external threads can be created in SolidWorks, depending upon the edge that is selected. If the circular edge of a hole is selected, then an internal thread is created. If the edge of a cylinder is selected, then an external thread is created.

To create a cosmetic thread, select the circular feature on the drawing that is to be threaded. Then, use **Insert** ⇒ **Annotations** ⇒ **Cosmetic Thread** or **Cosmetic Thread** in the **Annotations** toolbar. In the *Cosmetic Thread* dialog box, specify the depth of the thread, the major or minor diameter, and the **Thread Callout**. The **Thread Callout** is the specification for the thread. For instance, "1/4-20 UNC" indicates a thread with a major diameter of .25 inches, a pitch of 20 threads per inch, and the United Course standard thread series. Cosmetic threads can be placed in either parts or drawings.

For an internal thread, the hole should be the size of the tap drill diameter, and the major diameter should be the thread's nominal diameter. For an external thread, the cylinder should be the nominal diameter of the thread. The minor diameter should be the same as the tap drill diameter. Tap drill sizes and nominal diameters of standard threads can be found in any machinist's handbook.

PROFESSIONAL SUCCESS: GEOMETRIC DIMENSIONING AND TOLERANCING

Geometric Dimensioning and Tolerancing symbols can be added to parts, assemblies, and drawings. For instance, to specify a flatness form tolerance, select a face of a part, assembly, or drawing. Then click **Insert** ⇒ **Annotations** ⇒ **Geometric Tolerance**, or activate the **Geometric Tolerances** button in the **Annotations** toolbar. In the dialog box that appears, several items must be set in order to apply the geometric tolerance. Click on the **GCS** (Geometric Characteristic Symbol) button. Then select *Flatness*, followed by **OK**. Set the tolerance value in the ***Tolerance1*** field and click **OK**. The annotation that appears can be moved using the **Select** tool.

Geometric Dimensioning and Tolerancing symbols are one of many annotation items found in SolidWorks. Other examples include Notes, Cosmetic Threads, Surface Finish, and Datum Feature Symbols. Unlike dimensions, annotations have no bearing on the actual model of the part and are usually used to present more detailed information of the manufacturing process on the drawing.

10.4 CREATING A DRAWING OF THE PIZZA CUTTER ASSEMBLY

Assembly drawings are useful in showing how parts fit together to create the entire assembly and in labeling individual parts in an assembly. In the next few sections, you will create the assembly drawing of the pizza cutter, as shown in Figure 10.23. Drawings of assemblies are created in a manner similar to drawings created from parts. In this section, you will also create both a section view and a detail view of the pizza cutter assembly.

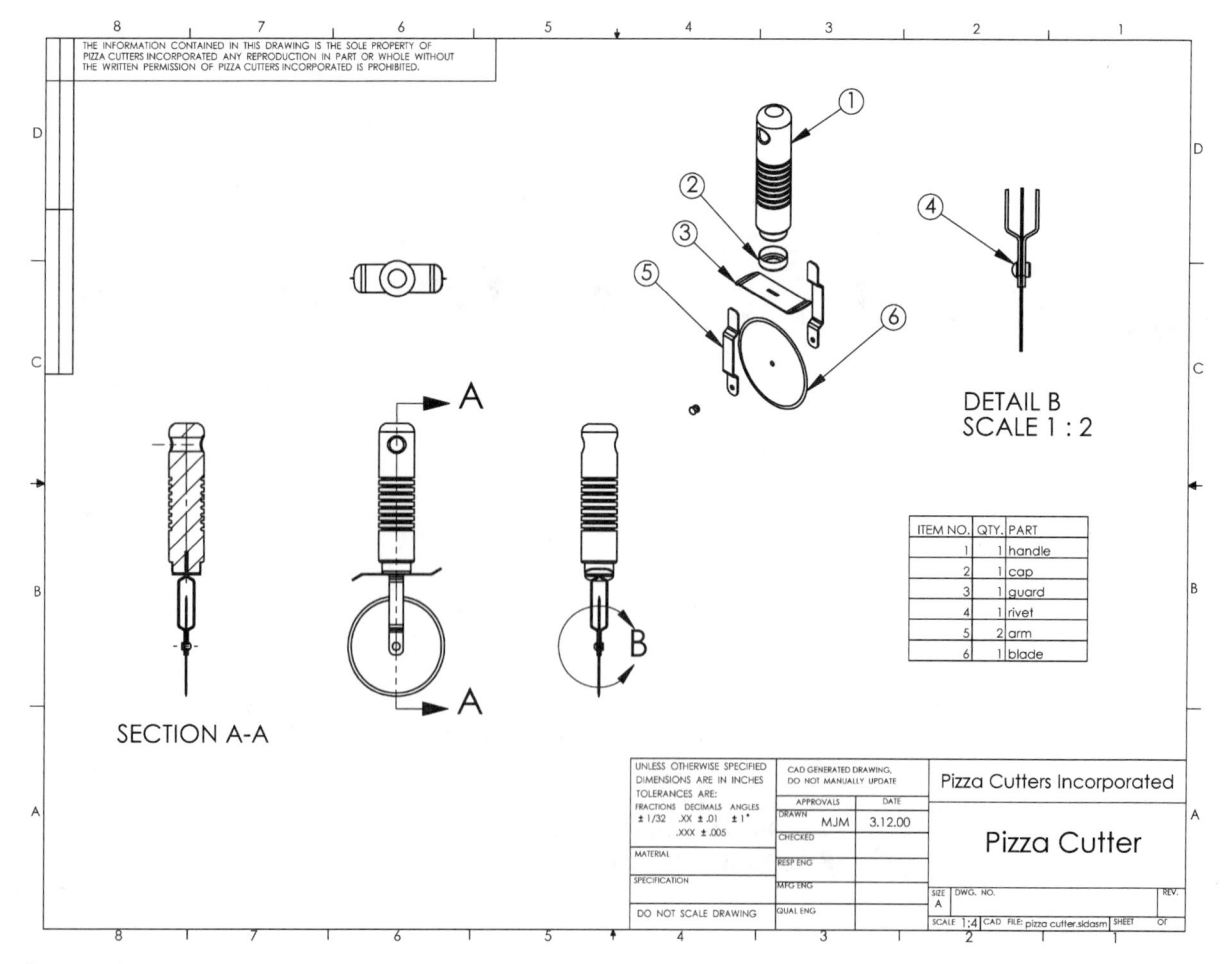

Figure 10.23. Finished assembly drawing

10.4.1 Setting Up the Pizza Cutter Drawing

1. Open a new drawing by clicking **<u>New</u>** in the **Standard** toolbar. Select **Drawing** from the dialog box. Click **OK**.
2. In the **Sheet Format To Use** dialog box, click on the **Custom Sheet Format** button. Click the **Browse** button, and open the file **tutorial format** that you saved earlier. Click **OK** in the **Sheet Format To Use** dialog box once you have selected the format. This opens a new drawing file. Normally, a larger drawing size should be used for this assembly drawing in order to show adequate detail. But to save the effort of making a new template *and* to allow printing on standard printers, you will use an *A* size drawing.
3. Click **Tools ⇒ Options**. In the **Document Properties** tab, activate the **Notes** item. Click the **Font** button and set the default note height to 8 points. Click **OK** twice.
4. Use the **Zoom To Area** tool to zoom in to the bottom right corner of the drawing. With the **Select** tool, *right click* on an open area on the drawing and select **Edit Sheet Format** from the menu, or select **Edit ⇒ Sheet Format**.
5. Remove the text "**18 GA STAINLESS TYPE 304**" and "**ASTM A240**" by selecting the text and hitting the delete key on the keyboard.
6. Change the text "**Arm**" and "**arm.sldprt**" to "**Pizza Cutter**" and "**pizza cutter.sldasm**", respectively. To do this, *right click* on the text, select **Properties**, and change it in the **Properties** dialog box. To reduce the size of the **pizza cutter.sldasm** to fit it into the field, deselect **Use Document's Font** in the **Properties** dialog box and change the font size to **6** points. Move the text to an appropriate position.
7. Now that the format is modified for the pizza cutter assembly, return to the sheet by *right clicking* on a open area of the drawing and selecting **Edit Sheet**, or by clicking **Edit ⇒ Sheet**. This permits the drawing to be edited instead of the format. **Zoom To Fit** to see the entire drawing.
8. Since this is an assembly drawing, you can set all of the views to **Hidden Lines Removed** to produce a drawing that is easier to read. Select **Tools ⇒ Options**. Activate the **Default Edge Display** item and set the **Default display of edges in new drawing views** to **Hidden removed**. Click **OK**.
9. Turn off **Origins** in the **View** menu.
10. In the **View ⇒ Toolbars** menu, be sure that **Standard**, **View**, **Annotation**, **Drawing**, **Sketch**, **Sketch Relations**, and **Sketch Tools** are checked.
11. *Right click* on any open space on the drawing and select **Properties**. Select **Third angle** and click **OK**.

10.4.2 Adding Orthographic and Isometric Views to the Drawing

Adding views of an assembly to a drawing is similar to adding views of a part to a drawing. You will be adding the front, top, right, and isometric views of the pizza cutter to this drawing. The isometric view will be exploded to show the assembly procedure for the pizza cutter.

1. Select **File ⇒ Open** to open the pizza cutter-assembly file. Find the file **pizza cutter**. Remember to set **Files of Type** to **Assembly Files** or **SolidWorks Files**, so that the file can be seen in the dialog box. SolidWorks will probably need to rebuild the assembly once it is opened. This is because of the modification to the arm's length that was made in the previous section. Click **Yes** to rebuild.

2. Hold down the Control key on the keyboard and press the Tab key once, to quickly switch back to the drawing window. Be sure that you are editing the sheet and not the sheet format in the drawing window. Click the **Standard 3 View** button in the **Drawing** toolbar. Then, return to the pizza cutter window using Ctrl-Tab. Select **pizza cutter** in the FeatureManager design tree to place the orthographic views into the drawing. Selecting **pizza cutter** in the FeatureManager design tree rather than in the Graphics Window ensures that the entire pizza cutter is selected, instead of a single part.

3. The *1:4* on the right side of the Status bar at the bottom of the screen signifies that the drawing of the pizza cutter will be one-quarter its full size. *Right click* on an open area and select **Edit Sheet Format**. *Right click* on an open area again and select **Annotations**, followed by **Note**. Type *1:4* for the scale and click **OK**. Move the **1:4** to the **SCALE** box next to **CAD FILE** box. It will be necessary to zoom in on the title block. *Right click* an open area, select **Edit Sheet** to return to the drawing, and **Zoom to Fit**.

4. Click the **Named View** button in the **Drawing** toolbar and insert an *Isometric* view into the drawing. Answer **Yes** to true dimensions.

5. *Right click* on the isometric view and select **Properties**. In the **Drawing View Properties** dialog box, check the *Show in exploded state* box to display the view as an exploded assembly. Click **OK**.

6. To make room for more views that will be added to this drawing, move the orthographic and isometric, as shown in Figure 10.24.

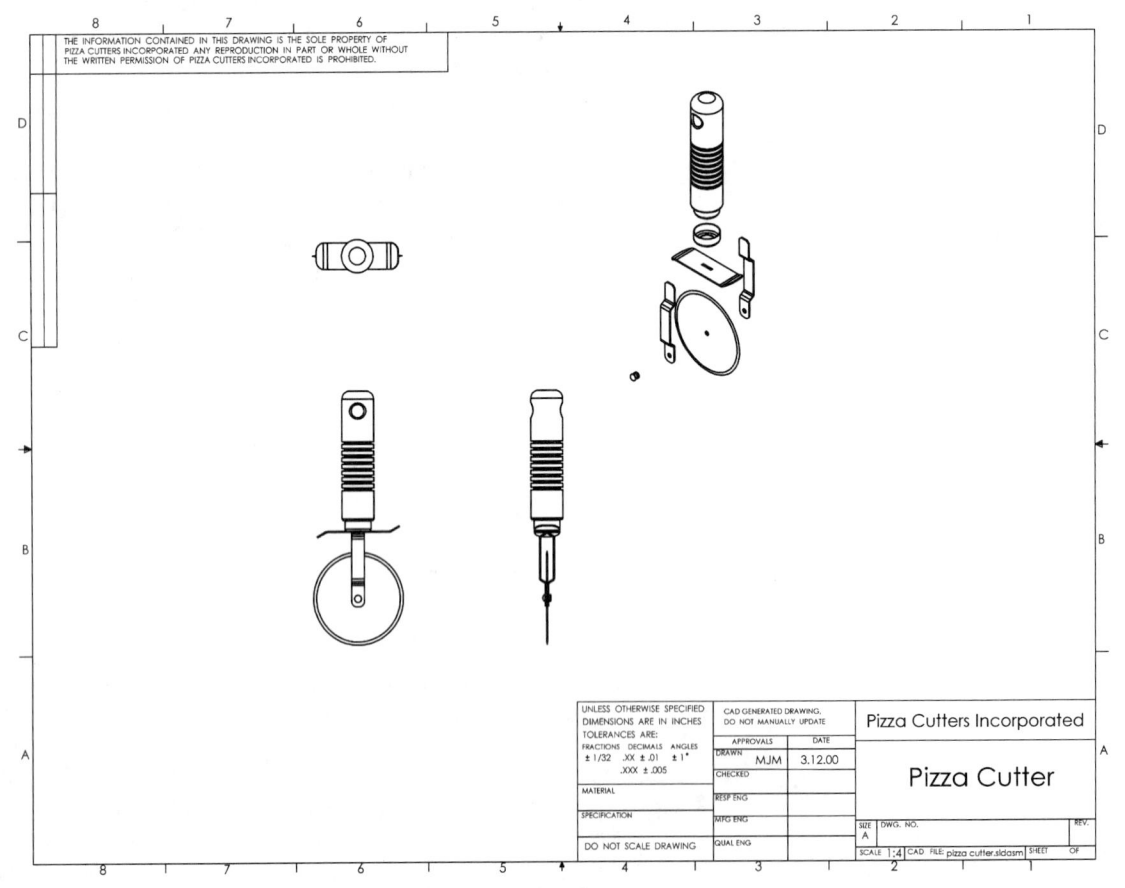

Figure 10.24. Orthographic and isometric views placed

10.4.3 Adding a Section View

Cross-section views are often useful to show how parts fit together. Cross sections can help avoid part interference in assemblies and ensure that interior regions of the parts are correct. "Slicing" a part or assembly along a plane results in a section view. A section view is created by sketching a line to indicate where the part or assembly is to be sliced and adding a new view of the sliced section. You will create a section view that slices the pizza cutter in half by cutting the front view of the model along a vertical plane.

1. Click on the front view to highlight it. **Zoom To Area** on the front view.

2. Select the **Line** tool in the **Sketch Tools** toolbar. Bring the cursor just below the lower edge of the blade in the front view of the pizza cutter. A light bulb will appear next to the cursor when the cursor is aligned with the centerline of the pizza cutter. With the light bulb visible, click and drag to draw a vertical line upward past the top of the handle. Look for the "V" next to the cursor before releasing the mouse button. The front view should look similar to the one shown in Figure 10.25.

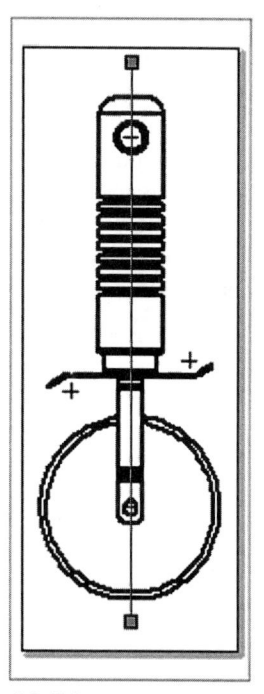

Figure 10.25. Line on the front view

3. Be sure that the line is selected (highlighted in green) and click the **Section View** button (two arrows pointing left) in the **Drawing** toolbar, or select **Insert** ⇒ **Drawing View** ⇒ **Section**. The *Section View* dialog box permits some parts of the assembly to be excluded from the section cut. This feature can make the section clearer. In this case, all of the components should be included in the section view, so click **OK** to the *Section View* dialog box. **Zoom To Fit** in order to see the entire drawing. Move the cursor toward the left side of the drawing and click once to place the section view. The section view appears, named **SECTION A-A**, as shown in Figure 10.26. Notice that the line that was drawn on the front view has changed to a section line with A's near the arrows, linking it to the section view. It may be necessary to zoom in to see these details. Move the views, if necessary.

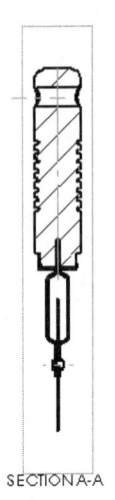

Figure 10.26. Section view

10.4.4 Adding a Detail View

A detail view is a enlarged view of a small portion of the drawing. A detail view can be particularly useful to show small details, such as the rivet of pizza cutter. The detail view is dependent on the parent view, but it has a scale that is independent of the rest of the drawing. This scale is usually larger than the rest of the drawing, so that the detail can be easily seen. To create a detail view, a circle or rectangle is sketched around the area to be shown in detail, and a new view of the enclosed area is added to the drawing. You will create a detail view of the lower portion of the pizza cutter, where the rivet holds the blade between the two arms, so that the detail of the rivet can be seen.

1. Click on the right view to highlight it.
2. Zoom in on the rivet in the right view. Activate the **Circle** tool in the **Sketch Tools** toolbar. Draw a circle centered on the rivet and similar in size to the circle shown in Figure 10.27. Although a circle was used in the Figure, any closed sketch could be used to create a detail view.

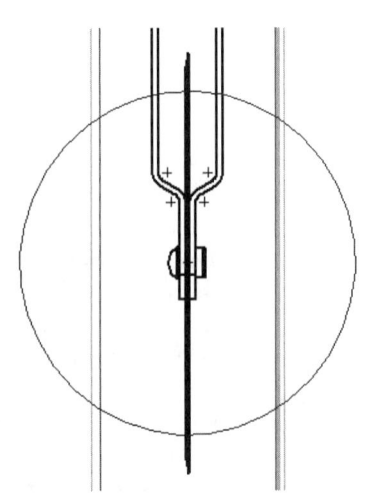

Figure 10.27. Circle for detail view

Section 10.4 Creating a Drawing of The Pizza Cutter Assembly   **211**

3. With the circle selected, click on the **Detail View** button (a circle with A in it) in the **Drawing** toolbar, or select **Insert** ⇒ **Drawing View** ⇒ **Detail**. **Zoom To Fit** and place the view in the upper right corner of the drawing. A detail view is placed with the name **DETAIL B**.
4. *Right click* on the border of the detail view and select **Properties** from the menu. You may need to zoom in to select the border.
5. Set the scale to *1:2* to enlarge the detail view. Click *OK*. The text "**SCALE 1:2**" appears below the detail view.
6. Set the font size to **16** points on both **DETAIL B SCALE 1:2** and **SECTION A-A** by *right clicking* on the text and changing the size in the **Properties** window. Click the **Redraw** button in the **Standard** toolbar to update the display. The drawing should look similar to that shown in Figure 10.28. Move the views to match this figure.

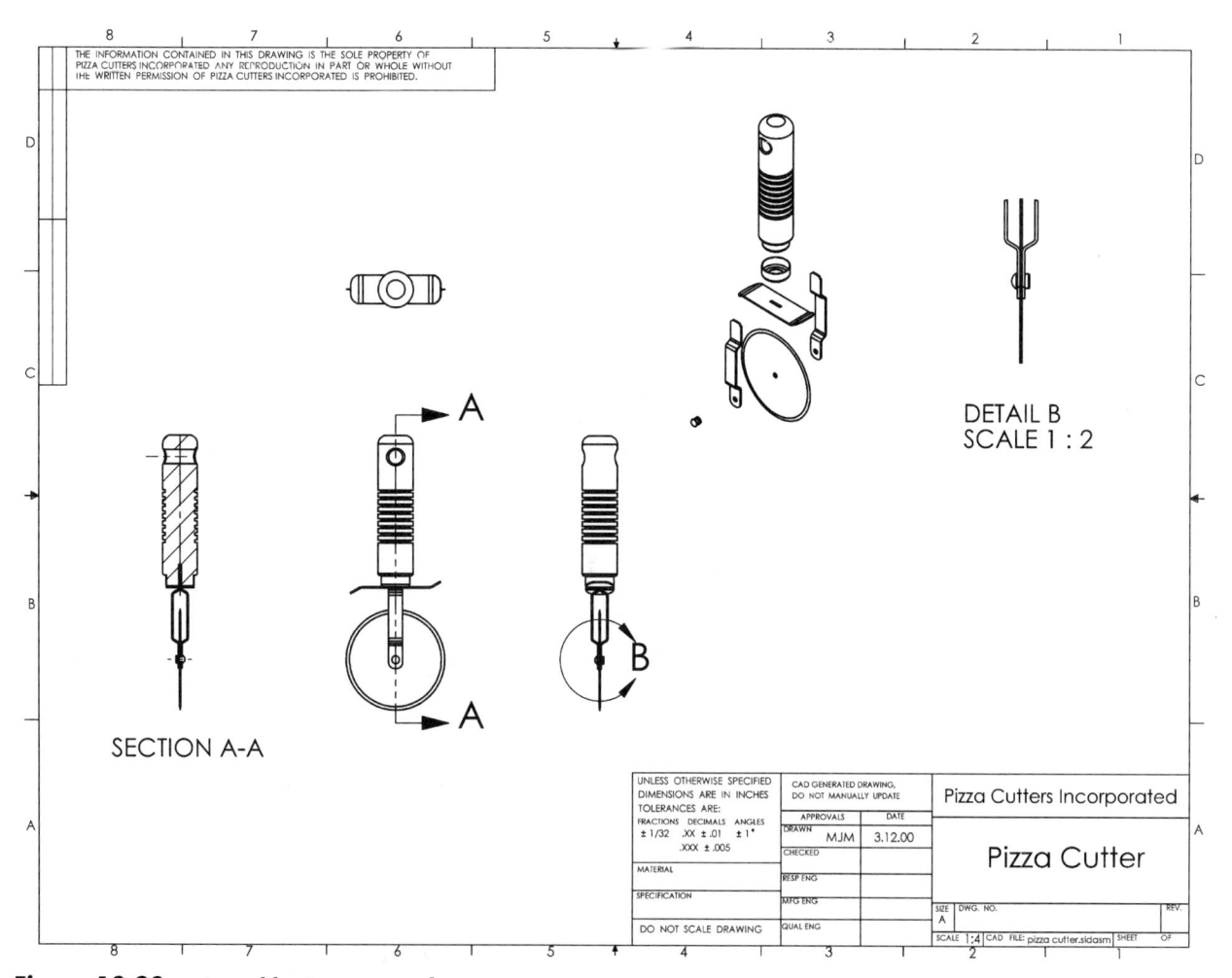

Figure 10.28. Assembly views arranged

10.4.5 Adding Numbers to the Components of the Assembly

In order to identify components in an assembly drawing, each component can be named or referenced by a number. In this section, numbers will be added to each of the parts of the assembly. A bill of materials referencing these numbers will be created in the next section.

1. Zoom in on the detail view. Click the **Balloon** button (a balloon with 1 inside) in the **Annotation** toolbar, or select **Insert ⇒ Annotations ⇒ Balloon**.

2. Click on the head of the rivet. An arrow and a number surrounded by a circle appear. It may be necessary to **Zoom To Fit**. The number refers to the item number SolidWorks assigned to the part in the assembly. If an arrowhead does not appear at the end of the leader line, *right click* on the number to activate the **Properties** dialog box. Turn off the **Smart** checkbox next to **Arrow Style**. Choose the filled arrowhead from the pulldown menu.

3. Change the font size of the balloon note to **14**. Click **OK** twice.

4. Click and drag the balloon note a short distance away from the blade, as shown in Figure 10.29.

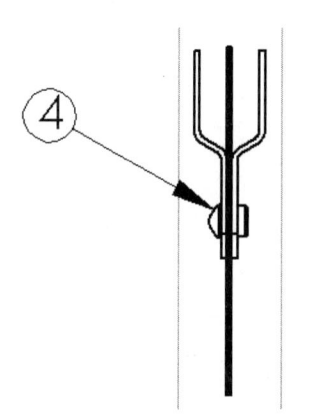

Figure 10.29.   Rivet numbered

5. Repeat Step 2 for the other parts of the assembly: guard, arm, blade, cap, and handle. Label the parts on the exploded isometric view. Move the labels to convenient locations, so that the arrows clearly point to the parts. The isometric view should look similar to the one shown in Figure 10.30. *Control click* on all of the labels to select them. Then *right click* on one label and change them all to **14**-point font size.

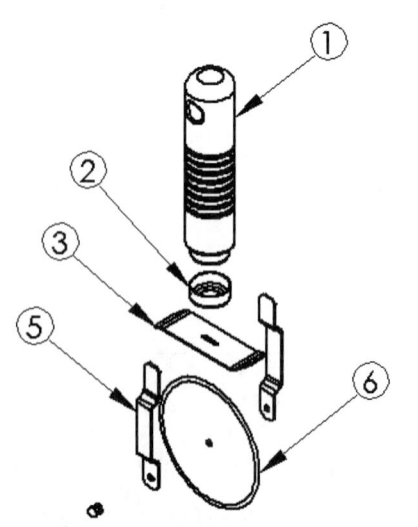

Figure 10.30.   Assembly components numbered

10.4.6 Adding a Bill of Materials

A bill of materials, or BOM, is a list of the parts in an assembly. Usually, a BOM shows the item number, the name, and the quantity of each item. If Microsoft Excel is installed on your computer, SolidWorks can automatically generate a BOM for you. In this section, you will create a BOM of the parts in the pizza cutter assembly.

1. Select the isometric view by clicking on the view within the box, but not on the pizza cutter. The view becomes highlighted in green. You can select any view, as long as all of the parts that you want on the bill of materials are present in it.

2. Select **Insert ⇒ Bill of Materials**. The *Select BOM Template* dialog box appears. Open *bomtemp*, a standard SolidWorks template for the bill of materials. The *Bill of Materials Properties* dialog box that appears is shown in Figure 10.31. Be sure that *Use the document's note font when creating the table* is checked, so that the font for the bill of materials matches the font of the rest of the drawing. Also, be sure that the **Show Parts Only** button is checked. With this button checked, all parts of the pizza cutter assembly will appear in the bill of materials, but the cutter sub-assembly will not. Deselect *Use table anchor point*, so that you can drag the bill of materials to any part of the drawing.

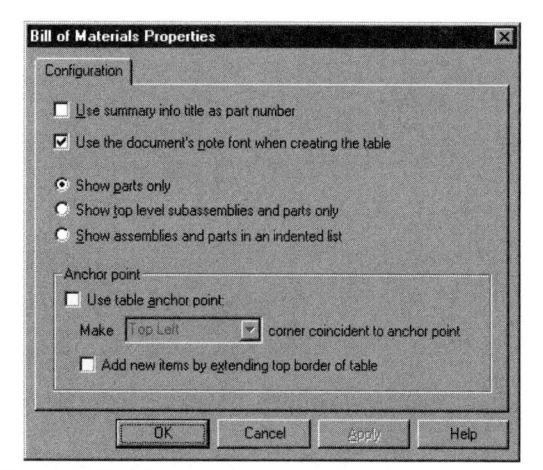

Figure 10.31. Bill of Materials Properties dialog box

3. Click **OK**. After a short wait, SolidWorks generates an embedded Excel table in the drawing. Click and drag the table to the right side of the drawing, as shown in Figure 10.32. You may need to move some views in order to make room for the BOM.

4. Zoom in on the table. There are four columns, with a row for each of the parts in the assembly. Notice that the numbers created in the balloon notes refer to the same parts in the bill of materials. SolidWorks automatically put a **2** in the **QTY.** column for the arm, signifying that there are two identical arms in the assembly. Activate the **Select** tool.

214 Chapter 10 Creating Working Drawings

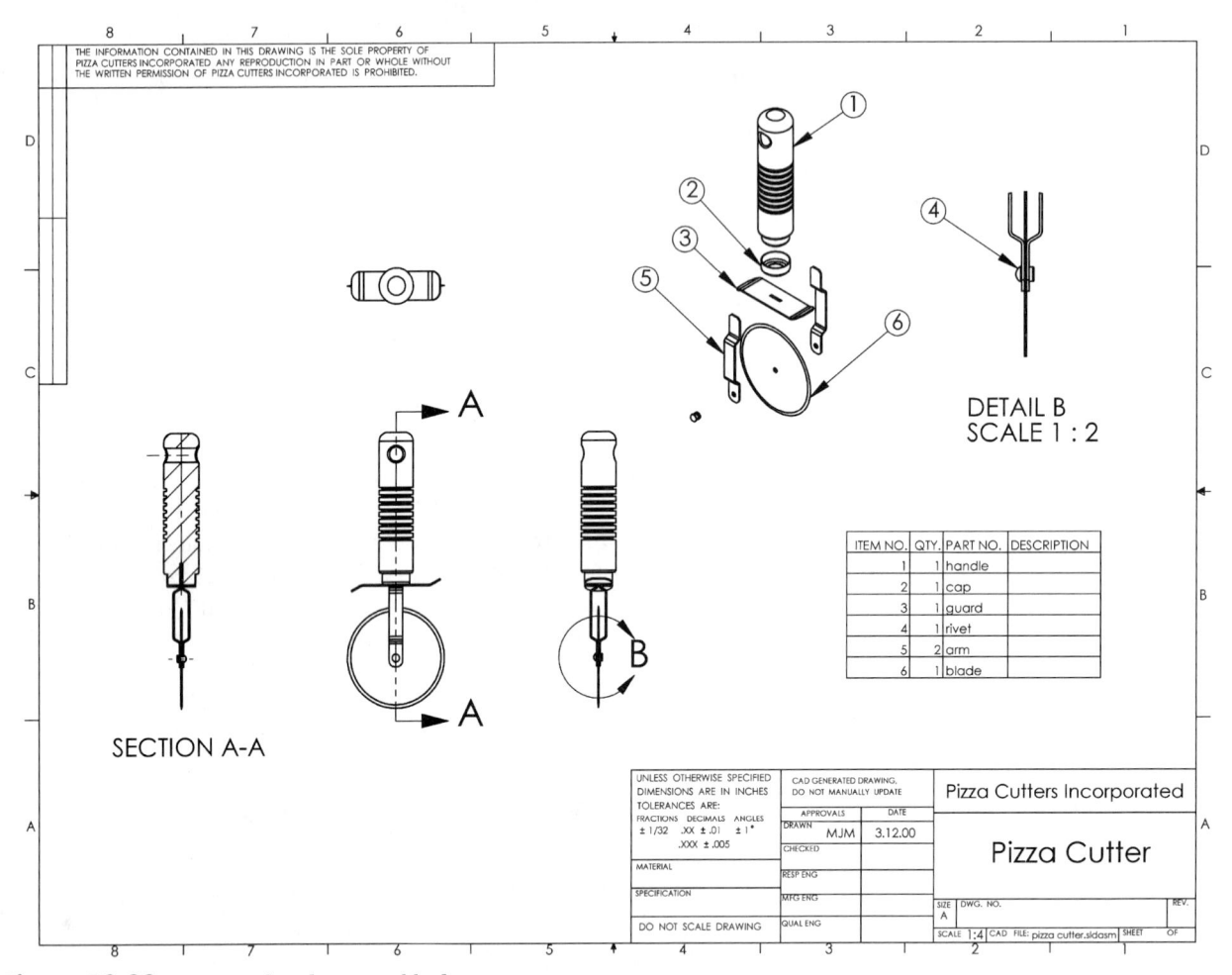

Figure 10.32. BOM placed in assembly drawing

5. The bill of materials can be edited by *right clicking* on the table and selecting **Edit Bill of Materials**. An editable Excel table appears, similar to the one shown in Figure 10.33. You may need to zoom out in order to see the embedded table. *Right click* on "**D**" at the top of the fourth column in the Excel window, and select **Hide** from the menu. This removes the Description column.

Figure 10.33. BOM Excel edit window

6. Change the text "**PART NO.**" to "**PART**" by clicking on the field and editing the text at the top of the screen, just as you would in Excel. Hit Enter to accept the change. Click on the SolidWorks portion of the screen to return to the drawing.
7. **Zoom To Fit** and move the items on the drawing to match Figure 10.23.

Congratulations! This completes the drawing of the pizza cutter assembly. Save your work as *pizza cutter drawing*. Check the drawing in **Print Preview**, make any necessary changes, and print the drawing using **File ⇒ Print**.

SolidWorks has many more capabilities than those covered in this tutorial. Now you have the basics necessary to model fairly complex parts. As you become more proficient with SolidWorks, you will be able to explore the advanced capabilities of the software.

PROFESSIONAL SUCCESS: FILE TRANSFER BETWEEN CAD SYSTEMS

As the engineering world becomes more interconnected with e-mail and the Internet, engineering graphics are routinely transferred as electronic documents both within and outside of a company. Sometimes, this involves the necessity of importing drawings created using one CAD system to a different CAD system. Each CAD system usually saves its files in its own format. Fortunately, several standard formats exist and can be used to transfer documents between different CAD programs. Translators are usually available within most CAD software to import and export data files that are in a standard format.

The most commonly used format for solids modeling applications is the Initial Graphics Exchange Specification, or IGES. Models may be imported to or exported from most CAD systems if the file is stored as an IGES transfer file (.igs). Unfortunately, IGES translators vary quite a bit and may sometimes give unexpected results, especially for the inexperienced user. Nevertheless, IGES is frequently used. Other standard file formats are available. STEP is capable of translating a solids model and maintaining it as such. DXF and DWG formats are intended for two-dimensional data, such as drawings.

SolidWorks can open a variety of file formats. In the **Open** dialog box, select the file format in the **Files of type** menu. Files can be saved in standard formats using the **Save as type** menu in the **Save As** dialog box.

KEY TERMS

Associative modeling
Bill of materials
Detail view
Drawing size
General tolerances

Isometric view
Orthographic view
Principal view
Projection
Scale

Section view
Sheet format
Tangent edges
Title block

Drawing Problems

1. Create a drawing of the guard. Include three orthographic views and an isometric view.
2. Create a drawing of the cap. Include a section view through the cap, in addition to two orthographic views and an isometric view.
3. Create a drawing of the blade. Include a detail section view of the cutting edge, in addition to two orthographic views and an isometric view.
4. Create a drawing of the rivet using a scale of 5:1. Include a section view through the rivet, in addition to two orthographic views and an isometric view. (Note that it is unlikely that a drawing of the rivet after it is deformed would be necessary in a practical situation.)
5. Create a half-scale drawing of the handle. Include a section view through the diameter of the handle and a detail section view of the rectangular cut at one end of the handle, in addition to two orthographic views and an isometric view.

Index

A

Actual size 68
Add Relation 111
Align Parallel/Concentric 191
Aligned Section View 190
American National Standards Institute 50
Angles 54
Angular dimension 98
Angularity 77
Annotation 191
Arcs 146
Assembly 155, 157, 159
Assembly drawings 18, 61, 206
Assembly files 86
Assembly sections 62
Associative 45, 186, 204
Automatic Solve 150
Auxiliary View 16, 190
Axis 136
Axis of revolution 127

B

Balloon 191, 212
Base feature 43, 91, 101
Basic size 68
Bilateral tolerance 69
Bill of materials 19, 213
Blind 119
Blind hole 60
Block 191
Blocking-in 26
Bolts 61
Boring 71
break line 52, 58
Broaching 71
Broken-out section 58

C

CAD 6, 24, 37
Center Mark 191, 202
Centerline 52, 55, 94, 127
Centerpoint Arc 94, 146

Chain dimensioning 70
Chamfer 114, 120
Child 45, 122
Chrysler Walter P. 1
Circle 25, 30, 54, 94, 104
Circular Sketch Step and Repeat 95
Circularity 76
Close 108
CNC machine tool 42
Coincident 161, 162
Collinear 135
Color 143
Computer Aided Design 6
Computer Aided Design and Drafting 6
Computer Aided Drafting 6
Computer Aided Engineering 6
Concentric 160
Concentricity 77
Conceptual design 7
Concurrent engineering 7
ConfigurationManager 183
Constrain 96
Constraint-based modeling 43, 95
Constraint-based solids modeling 41
Construction geometry 127
Construction lines 26
Constructive solid geometry 40
Contour dimensioning 53
Convert Entities 94
Cosmetic Thread 191
Crest 60
Crop View 191
Cross-section 41, 43, 91
Cross-section views 209
Cut 105
Cutting plane 56
Cutting plane lines 52, 56
Cylindricity 76

D

da Vinci Leonardo 2, 3
Datum dimensioning 70
Datum Feature Symbol 191
Datum plane symbol 77
Datum planes 91

Datum Target 191
Delete 97
Descartes René 4
Descriptive geometry 5
Design intent 43, 45
Design process 7, 17, 23
Design specification 7
Detail drawings 18
Detail view 59, 62, 190, 210
Diameter 54, 61
Dimension 92, 96
Dimension lines 52
Dimension placement 52
Dimensioning 55
Dimensions 52, 97, 98, 200, 202, 204
Dimetric projection 13
Drawing 185, 190
Drawing files 86
Drawing sheet format 186
Drawing size 187
Drill sizes 72
Drilling 71
Dürer Albrecht 4
DWG 215
DXF 215

E

Edit 175
Engineering Graphics 3
Engineering graphics 2
Exploded view 3
Extend 94
Extension 86
Extension lines 52, 55
External thread 60
Extruded 38, 41, 43, 91, 100
Extruded Boss/Base 101
Extruded Cut 105

F

Farish William 5
Fasteners 59
Feature control box 76

Index

Feature symbol 75
Feature-based modeling 43
FeatureManager design tree 90, 116, 120, 121, 158, 160
Features 43, 89, 90
File 90
File Transfer 215
Files 86
Fillet 94, 100, 106, 132
Filter Faces 156
Finite Element Analysis 42
first-angle projection 16, 49
Flatness 76
Forging 73
Form tolerances 76
Format 190, 191
Freehand sketches 7, 17
Freehand sketching 13, 23, 24
Full section 56
Fully Defined 96, 98

G

Gages 67
General tolerances 69, 194
Geometric Dimensioning and Tolerancing 74, 75, 191
Glass box 14
Graph paper 30
Graphics Window 87
Grid 92
Grinding 71

H

Half section 57
Help 90
Hidden In Gray 103
Hidden lines 39, 51
Hidden Lines Removed 103
Hide 138
Hide Component 169
History-based modeling 45
Hole Callout 191
Hole Wizard 108
Holes 108
Honing 71

I

Ideation 7, 17, 23
IGES 215
Inch dimensions 53
Integrated product and process design 7

Interchangeable parts 68
Interference 174
Internal thread 60
International Standards Organization 50
Intersection curve 94
Isometric freehand sketches 27, 30
Isometric grid paper 32
Isometric projection 12, 26
Isometric view 5, 18, 103, 186, 197

J

Jigs 68

L

Landscape 188
Lapping 71
Lay 73
Leader lines 52, 55
Least material condition 80
Limit dimensions 69
Line 94, 95
Linear Pattern 137
Linear Sketch Step and Repeat 95
Lofts 153

M

Major diameter 60
Manufacturing process 71
Mate 160, 161
Maximum material condition 79
Menu bar 87
Metric threads 61
Millimeter dimensions 53
Milling 71
Mind's eye 1, 23
Minor diameter 60
Mirror Feature 121
Mirroring 121
Model 37
Model Items 191
Modify 99
Monge Gaspard 4
Move Component 159, 161
Multiple features 55
Multiview projections 11, 14

N

Named View 191, 197
New 90

Nominal 61
Nominal size 68
Normal To 103
Note 191
Nuts 61

O

Oblique freehand sketch 26, 28
Oblique projection 13, 26
Offset Entities 94
Offset section 58
Options 88
Orientation 102
Orientation tolerances 77
Origin 91
Orthographic projections 4, 14, 32, 186, 196
Orthographic views 18

P

Pan 103
Parallel 165
Parallelism 77
Parametric modeling 44
Parent 45, 122
Parent child relationships 45, 122
Parent view 210
Part files 86
Parts list 19
Pattern 134, 136
Permanent Mold Casting 73
Perpendicularity 77
Perspective 13
Photorealistic 26
Pictorial perspective 3, 13
Pitch 60
Plus and minus 69
Point 94
Polishing 73
Polygon 94
Positional tolerance zone 79
Positional tolerances 78
Precedence of lines 52
Primitive modeling 40
Principal view 198
Print 107, 205
Print Preview 205
Pro/ENGINEER 6, 42
Profile tolerances 76
Projected View 190
Projections 11, 12, 14, 198
PropertyManager 159
Punching 71

R

Radius 54
Rapid prototyping 42
Reaming 71
Rebuild 174, 205
Receding construction lines 27
Rectangle 94, 130
Redraw 99
Reference dimensions 202
Regenerates 99
Relation 110, 135
Relative View 191
Rendering 40
Revolve 41, 91, 128, 132
Robust design 68
Root 60
Rotate Component Around Axis 159
Rotate Component Around Centerpoint 159
Rotate View 103
Roughness 73
Roughness average 73
Roundness 76
Rounds 106
Runout 77

S

Sand Casting 73
Save 107
Sawing 73
Scale 51, 198
Schematic representation 60
Screw threads 60
Section View 56, 190, 209
Section-lining 56
Select 92
Selection Filter 156
Shaded 103
Shading 40
Sheet 190
Sheet format 207
Show Component 172

Sikorsky Igor 23
Simple Hole 108
Simplified representation 60
Single limit dimension 69
Sketch 41, 89, 91, 92
Sketch Entities 94
Sketch Relations 89
Sketch Tools 89, 94
SKETCHPAD 6
SmartMates 165
Snap 92
Solid feature 119
Solids modeling 6, 40, 42
SolidWorks 6, 42
Specification 7
Spline 94
Stacked Balloon 191
Stamping 71
Standard 89
Standard 3 View 190, 196
Standard size 50
Start 87
Status bar 87, 96
STEP 215
Straightness 76
Surface Finish 73, 191
Surface model 39
Sutherland Ivan 6
Sweeps 153
Symmetry 77
Symmetry lines 52

T

Tangent Arc 94, 146
Tangent edges 198
Target size 68
Text 202
Thin Feature 101
third-angle projection 15, 49, 195
Thread series 61
Title block 50, 186, 191
Tolerance 53, 68, 70, 203
Tolerance accumulation 70

Tolerance stackup 70
Tolerancing 68
Toolbar 89, 91, 94
Tools 88
ToolTips 92
Trim 94, 131
Trimetric projection 13
True-position dimensions 78
Turning 71

U

Uccello Paolo 3
Under Defined 96
Undo 102
Unilateral tolerance 69
Units 53, 92

V

View 89
View Orientation 102
Visible lines 51
Visual thinking 1

W

Waviness 73
Weld Symbol 191
Wireframe 103
Wireframe model 39
Witness lines 52, 151
Working drawings 17, 19

Z

Zoom In/Out 103
Zoom To Area 103
Zoom To Fit 103
Zoom To Selection 103